Recent Research in Psychology

Michael J. Renner   Mark R. Rosenzweig

Enriched and Impoverished Environments

Effects on Brain and Behavior

Springer-Verlag
New York Berlin Heidelberg
London Paris Tokyo

Michael J. Renner
Department of Psychology
University of Wisconsin, Oshkosh
Oshkosh, Wisconsin 54901, USA

Mark R. Rosenzweig
Department of Psychology
University of California, Berkeley
Berkeley, California 94720, USA

With two Illustrations

Library of Congress Cataloging in Publication Data
Renner, Michael J.
Enriched and impoverished environments.
(Recent research in psychology)
Bibliography: p.
Includes index.
1. Brain—Adaptation. 2. Neuroplasticity.
3. Neuropsychology. I. Rosenzweig, Mark. R.
II. Title. III. Series. [DNLM: 1. Behavior—
physiology. 2. Brain—physiology. 3. Environment.
WL 300 R414e]
QP376.R45 1987 612'.82 87-4925

Printed and bound by Quinn-Woodbine, Woodbine, New Jersey
Printed in the United States of America.

9 8 7 6 5 4 3 2 1

ISBN 0-387-96523-8 Springer-Verlag New York Berlin Heidelberg
ISBN 3-540-96523-8 Springer-Verlag Berlin Heidelberg New York

To
Dr. John W. Renner and Carol J. Renner
To
Janine S. A. Rosenzweig

Acknowledgements

The authors are grateful for support from several sources. The University of California has been consistently supportive of work on the environmental influences on brain and behavior over the years, including grants from The University of California Chancellor's Patent Fund and other internal sources. This research has received support also from several sources within the United States federal government, including the National Science Foundation, the Departments of Education and Energy, and several agencies of the United States Public Health Service.

We would also like to extend our thanks to the several colleagues who provided valuable feedback on the manuscript, and to Ms. Kathy Ludwig for her assistance with the references section of this monograph.

Table of Contents

Chapter 1:
Introduction and General Overview

A recent newspaper advertisement showed a large photograph of a human brain and underneath it the boldface caption, "INCREASE THE SIZE OF YOUR ORGAN" (*San Francisco Examiner*, September 13, 1986, page B26). The accompanying text urged readers to subscribe to the paper in order to exercise their brains and thus increase the size and capacity of this organ. Apparently the advertisers think that the public is willing to entertain, at least speculatively, the idea that stimulating the brain has beneficial effects, both anatomically and intellectually.

The idea that exercise can increase the size of the brain and mental prowess has had its ups and downs over the last two centuries. Perhaps the first to propose the hypothesis that neural tissue might respond to exercise by physical growth (in a like fashion to the response of muscles) was the Swiss naturalist Charles Bonnet (1720-1791), in his correspondence with the Italian scientist Michele Vincenzo Malacarne (1744-1816). Malacarne, however, was the first to conduct experimental work concerning this hypothesis: In 1791 Malacarne reported an experiment in which he had divided birds from the same clutches of eggs into two groups; one group was given experience and the other was kept in isolation. At the end of the experimental period, Malacarne examined the brains of the birds and found that those given enriched experience were larger than those of the isolated birds, especially in the cerebellum. The well-known physiologist Samuel Thomas von Soemmering (1755-1830) undoubtedly was referring to this report when he wrote in 1791 that anatomical measurements might demonstrate the effects of experience on the brain. By the time his book was revised in 1800, however, Soemmering had apparently developed new reservations. In this edition, he added the qualification "although anatomy has not yet demonstrated this" to his analogy between exercise of muscles and of brain tissue (p. 394).

In the latter part of the nineteenth century, several investigators wrote about effects of experience on the brain and on ability. Thus Charles Darwin (1859) wrote that domestication reduced brain size; that is, relative impoverishment of experience in the domestic setting led to reduced development of the brain. Toward the end of the last century a self-styled professor, scientist, and inventor named Elmer Gates claimed to have conducted animal research that supported his hypothesis of "brain building," that is, "that every conscious mental operation or experience creates in some part of the brain or nervous system new structural changes of cell and fiber... [producing] the embodiment of more mind" (1909). *Who's Who in America* for 1918-19 listed Gates as a psychologist and scientist and stated that he had "Evolved a practical art of brain or mind-building by systematic means, which causes an increase in the structural elements of the brain-cells, fibers and whole nervous system, increases mental capacity and skill" (p. 1017). Apparently Gates did not publish this research in scientific journals, but some of his articles appeared in *The Metaphysical Magazine* and were then collected by the Theosophical Society in a book in 1904. The first chapter of the book starts with a description of research quite similar to that of Malacarne; we will also see that Gates did not hesitate to draw upon his results to prescribe for human education:

> "The first experiment in my investigations ... consisted in giving certain animals an extraordinary and

excessive in one mental faculty -- e.g., seeing or hearing -- and in depriving other animals, identical in age and breed, of the opportunity of using that faculty. I then killed both classes of animals and examined their brains to see if any structural difference had been caused by excessive mental activity, as compared with the deprivation or absence thereof. During five or six months, for five or six hours each day, I trained dogs in discriminating colors. The result was that upon examining the occipital areas of their brains I found a far greater number of brain-cells than any animal of like breed ever possessed.

"These experiments serve to localize mental functions, and, above all, to demonstrate the fact that more brains can be given to an animal, or a human being, in consequence of a better use of the mental faculties. The trained dogs were able to discriminate between seven shades of red and six or eight of green, besides manifesting in other ways more mental ability than any untrained dog.

"The application of these principals [sic] to human education is obvious ... Under usual circumstances and education, children develop less than ten per cent of the cells in their brain areas. By processes of brain building, however, more cells can be put in these otherwise fallow areas, the child thus acquiring a better brain and more power of mind... (pp. 9-10)."

An opposed trend was also taking place in the last quarter of the 19th century: Measurements of brain size and body size in health and disease were showing the relative constancy of the brain even when body size varied markedly. By the turn of the century it came to be accepted that brain size remains fixed after its full growth is attained. Convinced of this fixity, Ramon y Cajal made one the poorest guesses of his illustrious career. Believing that neurons must increase their ramifications as a result of training but also believing that brain size remains fixed in the adult, Ramon y Cajal proposed that the size of the neuronal somata must shrink to allow room for growth of the ramifications! (1894, p. 467). Actually, as we will see below, when the neural processes ramify in response to enriched experience or training, the somata also grow in order to support the increased metabolism of the extended branches.

As of the 1950's, the prevailing dogma held that the anatomy and physiology of the brain were fixed; development proceeded according to the genetic plan until the predetermined physical dimensions and chemical composition of the adult brain were reached. In the absence of trauma brought about by injury or disease, change in the adult brain ceased except for the inevitable decay brought about by the aging process. This doctrine of anatomical fixity of the adult brain was so thoroughly ingrained by the 1950s that when evidence was serendipitously obtained that training or enriching experience lead to growth of the cerebral cortex, the experimenters did not even recognize the difference at first (Krech, Rosenzweig, and Bennett, 1960). This monograph recounts how the effects of differential experience on brain and behavior have been studied in the quarter century since the first demonstration (Rosenzweig, Krech, Bennett, & Diamond, 1962), with emphasis on the more recent developments and findings in this field.

The Basic Phenomenon

The fundamental empirical finding is as follows: Rats placed in a complex, challenging environment (designated an "enriched condition," or EC) develop reliable differences in some brain measures and some aspects of behavior from littermates placed in a relatively less stimulating environment (designated as an "impoverished condition," or IC). Neurobiological differences between animals with enriched and impoverished experience are discussed in detail in Chapter 2;

behavioral effects of differential experience are covered more fully in Chapter 3.

The animals that are placed into different stimulus environments are in most experiments genetically similar or identical. Any differences in biology and behavior observed subsequently can therefore be attributed to environmental influences.

Dimensions of Differential Experience

The environments used in the laboratory and described as enriched were originally adaptations of the "free environments" described by Hebb (1947). Typically, a group of 10 to 12 animals are placed in a relatively large cage (75 x 75 x 40 cm in many studies with rats) with a number of junk objects such as metal and cardboard tubes, wooden blocks, metal ladders and chains, and other objects. A detailed descriptive list and photograph of typical objects can be found in Rosenzweig and Bennett (1969), although almost any type of object can serve the purpose of enriching the stimulus complexity of the cage environment. On a relatively frequent basis, some of the objects in the environment are removed and replaced with others from a collection of objects kept available in the laboratory for this purpose.

The impoverished environment consists of standard laboratory cages, in which animals are housed singly. (It is interesting to note that, with increasingly stringent regulations in the United States concerning animal care, this type of housing is increasingly used as the standard laboratory housing condition.)

Animals in all conditions in nearly every experiment described in this monograph are housed in stable temperature conditions under normal day-night light cycles, and with ready access to ample food and water. Exceptions to this generalization, quite few in number, are noted in the text as the studies containing the exceptions are described.

Figure 1 shows relatively typical examples of enriched and impoverished environments; these are provided in order that mention of variations from the most common procedures can be set in proper context.

The terms enriched and impoverished are clearly relative. The complexity of an animal's environment can vary greatly, from extremely impoverished conditions (e.g., sensory deprivation) at one end of a hypothetical continuum to dangerous and unpredictably complex conditions at the other end. When set into this context, it is apparent that the manipulations employed in the laboratory to create environmental differences are actually quite moderate, creating experiences that obviously differ, but are not as different in absolute terms as would be possible.

The relative nature of the term enrichment does, however, lead to semantic complications when discussing and comparing experimental methodologies. Environments that have been described without qualification as "enriched" have varied widely. An example of weak enrichment was a small cage housing two subjects for 30 days with a jello mold and a single angle iron fixed to its sides (McCall, Lester, and Dolan, 1969). An example of considerable enrichment was a 72 square meter multi-compartment enclosure, strewn with junk objects, criss-crossed with tunnels, and exposed to climatic variations (Holson, 1986). There is in fact some evidence that the degree of environmental complexity is correlated with brain measures (Renner, Rosenzweig, Bennett, and Alberti, 1981; see also Chapter 4 of this monograph). Obviously, without some metric for the degree of environmental complexity, it makes little sense to speak in general terms across experiments, using procedures as widely disparate as those described above, of the effects of "complexity" or "enrichment."

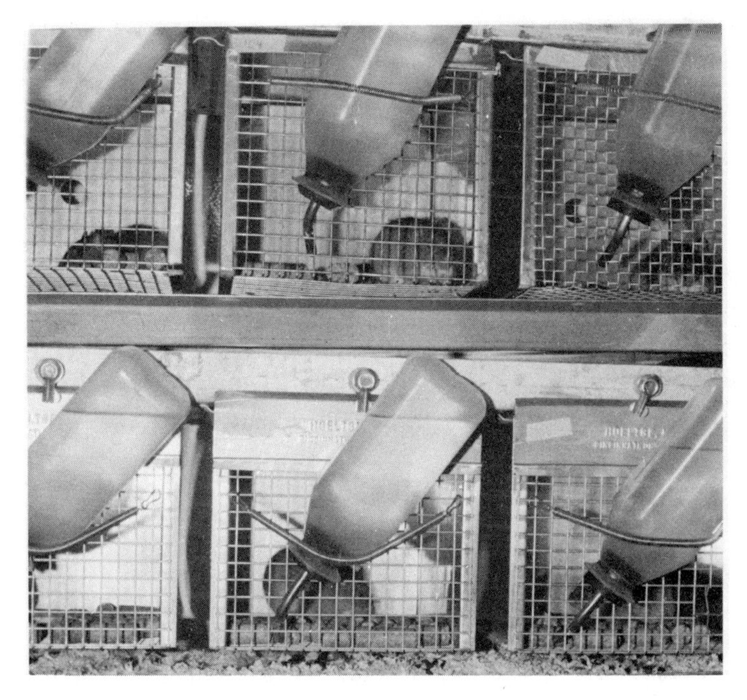

Figure 1: Enriched (top) and impoverished (bottom) environments.

Until and unless the definitive experiment is performed wherein environmental differences are appropriately scaled for their contribution to brain and behavioral differences from impoverished animals, the definition of an "enriched" environment must necessarily be somewhat arbitrary. In this monograph, we will discuss studies as having employed an enriched condition when the laboratory environment thus labelled contains, at a minimum, (a) some form of socially grouped housing and (b) considerable opportunity for physical interaction with inanimate stimuli, such as by including an elaborate spatial layout, multiple objects with which the subject may interact, or subjecting some aspect of the environment (available for direct investigation) to relatively frequent change.

Some of the experiments reported in this monograph included additional environmental conditions. These usually take the form of some middle ground between IC and EC. For example, animals are sometimes included in analyses after having been housed in "social conditions" (SC), with a small number of animals housed together, or in "grouped conditions" (GC), in the same type of cages as are used in the EC, but without stimulus objects. In general, brain and behavioral measures from animals thus housed are between those from EC and IC animals; there is modification in both directions from SC or GC subjects. In view of this bidirectional modification, it is not accurate to classify EC-IC effects as due solely to deprivation or enrichment alone. For the sake of clarity in the presentation, however, the data from SC and GC groups will be omitted from the discussions that follow, unless those data contradict the pattern that brain and behavioral measures of SC or GC are between EC and IC values.

Purpose and Goals of This Monograph

The question of how experience is recorded is fundamental to psychology; speculations and investigations concerning the role of the brain in this process have entered a particularly exciting phase as of the late 1980's. A major subfield in the neurosciences, presumed by most investigators to play a central role in storage and processing of information in the brain, is the study of plasticity in the nervous system at both molar and molecular levels. Studies of the effects of manipulating the external environment are an important part of this area of research, and there is a periodic need to organize and reflect upon what is known and what is yet to be known from studies of this type.

Manipulation of environmental complexity was one of the earliest methods utilized in the study of neural plasticity; as techniques for study of the nervous systems become capable of providing ever more detailed information concerning the modifiability of specific aspects of neural function, a continuing stream of new reports from numerous investigators demonstrates that studies of the response of major components of the nervous system to changes in a naturalistic environment continue to provide important information. This information comes by several routes: (a) by identifying particular parameters of neural anatomy and neurochemical function that are susceptible to experiential alteration and therefore may be fruitful for more detailed study, (b) by documenting regional variations in responsiveness to environmental change and providing clues about localization of neural functions, and perhaps most importantly, (c) by providing an avenue for the study of brain-behavior relationships that allows consideration of both brain and behavior. Although the studies discussed here represent a different level of analysis than the highly molecular approach in vogue in the 1980's, they continue to yield important insights into the manner in which neurobiological systems adapt to input gained by experience with the external environment.

This monograph was written to serve two functions: First, to organize the evidence to date concerning the responsiveness of neural and behavioral systems to external manipulations of the general character of the environment. To fulfill this first function, we will consider the neuroanatomical, neurochemical, and behavioral consequences of differential experience, primarily through studies of rodents but also in other mammals where such evidence exists. We will also discuss some parallel findings from the study of formal training on brain measures in rodents. We have attempted to integrate more recent findings with earlier ones, in order to provide a document for reference that does not assume familiarity with the voluminous literature on studies of environmental enrichment and impoverishment.

Previous reviews of this field include those of Greenough (1976), Rosenzweig and Bennett (1977, 1978), and Walsh (1981). This is, however, an active area of research; in the six years since the appearance of a comprehensive review of this area, substantial advances have occurred, both in empirical knowledge and in understanding of the effects of differential experience. This monograph, which is far from exhaustive, cites 102 such studies dated 1980 or later, coming from over three dozen different laboratories.

The second function of this monograph is to consider more fully the issues of causation of the effects of environment on brain and behavior, by addressing the plausible hypotheses concerning causes and the experimental evidence relevant to each. We hope that this monograph will provide tools and encouragement for further study, and that it will prompt the formulation of new questions and hypotheses.

In the following chapter, we will discuss the wealth of data concerning biological differences between animals after experience in enriched and impoverished environments. Chapter 3 will review the evidence concerning behavioral differences between animals from these experiential backgrounds. We will then take up in Chapter 4 the generalizability of these effects, reviewing studies using different strains, species and sexes in these experimental settings. Chapter 4 will also address the limitations on environmentally induced neural change, by examination of studies concerned with the time required for the appearance of these effects and the maximum magnitude of differences inducible by environmental manipulation. Chapter 5 addresses the critical question of cause: What biopsychological mechanisms can be identified as responsible for the differences we observe, and what behavioral events create the cerebral climate that leads to the changes making the effects of experience evident in the brain? Chapter 6 brings us to the existing knowledge and further potential for the beneficial application of findings from the literature on laboratory studies of enriched and impoverished environments.

Chapter 2:
The Neurobiology of Differential Experience

The biology of the organism is demonstrably altered by experience. Environmentally impoverished rats consistently outweigh their littermates from the enriched environment (Bennett, Diamond, Krech, and Rosenzweig, 1964); this co-occurs with increases in several measures of skeletal size, including external dimensions of the skull but not including intracranial capacity (Diamond, Rosenzweig, and Krech, 1965). These results are at least partially explained by the fact that rats with impoverished experience are less active and eat more (Fiala, Snow, and Greenough, 1977). Cummins and Walsh (unpublished data, discussed in Walsh, 1980) report increases in several organs, including liver, spleen, testicles, and heart, corresponding to those obtained with body weight. These weight differences are not, however, indications of differences in the general health of the animals.

The most behaviorally meaningful biological changes resulting from differential experience, however, are found in the central nervous system. The brain is affected both structurally and chemically: although it is somewhat arbitrary to separate any aspect of neurobiology into its anatomy and chemistry, we begin with such a division, perhaps sacrificing a bit of conceptual purity for clarity of organization.

Neuroanatomical Changes in Response to Differential Experience

Animals placed in differential environments show alterations in the structural characteristics of many aspects of the central nervous system. Manipulation of the stimulus environment has led to significant anatomical modifications, seen at many levels of observation, including changes in gross weight of the brain, weight and thickness of the cerebral cortex, microscopic changes in cell density and relative proportions of different cell types, and changes in the structure of individual neurons. Each of these will be discussed below. As the majority of differential environment studies have been conducted on structure and function of the neocortex, these studies shall form the primary focus of the sections to follow. Studies examining specific structures outside the neocortex will appear at the end of the section on neuroanatomical effects of differential environments.

Weight of Neocortex and Neocortical Regions

The most obvious gross anatomical change is an increase in total cortical weight that may reach about five percent (Rosenzweig, Bennett, and Diamond, 1972b, 1972c). Although this effect is not large in a quantitative sense, a change of this magnitude in such a major component of the central nervous system could easily exert an important influence on behavior. For example, ablations in the occipital neocortex, involving less than five percent of total cortical tissue, can render an animal functionally blind. Similar lesions, caused by stroke or focal head trauma, can cause aphasia or other disorders in humans.

This five percent gross difference in cortical weight is not a product of uniform increases in weight across the cortex. In examining regional specificity of brain responses to differential environments, many experiments have used a standardized dissection procedure, described in detail by Bennett and Rosenzweig (1981), and used in their laboratory since the 1960s. It was based on a functional map of the rat cerebral cortex developed by Zubeck (1951). This procedure allows samples of the various functional areas of the neocortex to be taken in a reliable way. Using this technique, the regional aspects of environmental responses of cortical anatomy can be studied, permitting more detailed and precise descriptions than can be achieved with techniques that involve the total cerebral cortex; this in turn allows studies of the brain effects of environmental enrichment and impoverishment to provide clues concerning possible behavioral consequences and mechanisms of the brain's response to differential environments.

The largest magnitude of environmental effects is found in the occipital region of cortex; mean differences between EC and IC subjects are typically 8 or 9%. While this region of the rat cortex is electrophysiologically active during visual stimulation, the occipital cortex in the rat is best described as an intersensory (or polysensory) area. Rats blinded by enucleation show a shrinkage of occipital cortex, but they also show significant EC-IC brain effects, including effects in occipital cortex, of approximately the same percentage magnitude as sighted rats (Krech, Rosenzweig, and Bennett, 1963; Rosenzweig, Bennett, Diamond, Wu, Slagle & Saffran, 1969). The same basic finding is repeated in experiments conducted in total darkness: brain differences were found between enriched and impoverished rats of the same percentage magnitude as those found for rats in experiments conducted under standard laboratory lighting conditions (Rosenzweig, et al., 1969). Since it would not be reasonable that visual system plasticity would be elicited in animals with little or no elicited visual activity, these findings imply that the EC-IC differences in occipital cortex are not primarily visual in nature. Further, although there were preliminary reports that visual stimulation was responsible for enrichment effects (Singh, Johnston, and Klosterman, 1967, 1970; Singh, Johnston, and Maki, 1969; Singh, Maki, Johnston, and Klosterman, 1970), these findings were not replicable with other strains of rats (Maki, 1971).

Brown (1971) reported that "intensive visual stimulation" (IVS) produced effects on the cholinergic system (discussed below in a separate section) similar to those found in studies of enriched and impoverished environments; she used complex geometric forms hung from the ceiling and walls of a large group cage. Attempts to replicate these findings (Bennett, Rosenzweig, Diamond, Morimoto, and Hebert, 1974) however, revealed a cautionary phenomenon: the "visual" stimuli showed signs of wear and chewing, indicating that the rats in the IVS condition climbed upon and manipulated these objects, probably during the dark phase of the diurnal cycle, in a manner which converted the supposedly visual stimulation into a form of stimulation that is clearly multisensory, as is the more usual enriched condition.

Because the occipital cortex is where the effect of differential experience on gross cortical anatomy is the largest, the majority of studies that provide more detail concerning cortical responses to differential environments, cited below, are focused on samples of occipital cortex.

In addition to the occipital cortex sample, the rat cerebral cortex is subdivided into three regions by the dissection procedure mentioned above: (a) the somesthetic cortex sample, containing portions of the motor and sensory functional maps of the rat cerebral cortex; (b) the remaining dorsal cortex sample, containing parts of the visual and somesthetic areas (due to the way in which the boundaries of dissection samples are conservatively placed, so that the sample is taken from inside the boundaries of most rats' motor and sensory regions, portions of these areas are left to be included in other tissue samples) as well as the motor area and association areas of the rat; and (c) the ventral cortex, containing the hippocampus, amygdaloid complex, corpus callosum, and neocortical tissue ventral to the brain's widest point.

Differences in magnitude of cerebral effect among these three sections are not large, with the EC-IC comparisons typically yielding values for EC between 3-5% above those for IC in all three areas. Comparisons of cerebral effects in these areas are complicated by the variability of results from one study to the next: what emerges is not reliable within any given experiment, but is a statistically significant pattern of small differences in size of effect from one cortical region to another across numerous studies. Of these three cortical samples, the somesthetic cortex shows the smallest effect of environmental manipulation, with EC exceeding IC by approximately 3% over numerous experiments (Bennett, et al., 1964; Rosenzweig, Bennett, and Diamond, 1972c; Rosenzweig and Bennett, 1978). Differences between EC and IC groups for remaining dorsal cortex and ventral cortex are somewhat greater, with EC exceeding IC by an average of 4-5% in dorsal cortex, and by 3.5-4% in ventral cortex (Bennett, et al., 1964; Rosenzweig and Bennett, 1978).

Physical Dimensions of Brain Regions

Diamond, Krech and Rosenzweig (1964) reported that the thickness of the occipital cortex was 6.2% greater in rats housed in Environmental Complexity and Training (ECT) than in IC. A difference in thickness was found in all cortical layers except layer 1. (The enriched condition used in early experiments included daily training of subject in mazes. Unpublished experiments indicated that this training did not contribute substantially to EC-IC effects, and subsequent studies did not include this maze training.) In later work, these occipital effects were replicated and significant EC-IC thickness differences were found in somesthetic and motor cortices as well (Rosenzweig, Bennett and Diamond, 1972b).

Several laboratories have reported EC-IC differences in the dimensions of the cerebral cortex; these effects appear to be more dependent on time spent in the differential conditions than other measures. Altman, Wallace, Anderson, and Das (1968) first reported that 90 days in enriched conditions resulted in significant increases in cerebral length. With 30 days EC-IC housing, however, Rosenzweig and Bennett (1969) found nonsignificant differences in cerebral length and width in studies with both rats and gerbils (*Meriones unguiculatus*; see Chapter 4). Later investigations, however, provided a way of reconciling these results: Walsh, Budtz-Olsen, Torok, and Cummins (1971) reported that they obtained nonsignificant cerebral length differences of 1%, which were comparable to those obtained by Rosenzweig and Bennett, along with nonsignificant differences in cerebral width. When the length and width measures were combined by multiplication, however, EC-IC differences in the product (a crude estimate of neocortical area) were significant. Following 80 days' differential housing, differences in length and area measures had increased: The length difference was now significant, and the product of length and width remained so. The small differences in cortical length after 30 days differential housing, while smaller in magnitude than that found after 80 days, is reliable enough to produce significant results in larger groups (22 littermate pairs rather than the more typical 10-12; Walsh, Cummins, and Budtz-Olsen, 1973).

More recent reports have extended to the white matter the differences in size previously observed in the cortex. Szeligo (1977) found that the corpus callosum underlying the occipital cortex was thicker in enriched-experience rats, and that the number of axons was also greater in this fiber pathway than in rats with impoverished experience. Juraska and Meyer (1985) report similar differences in overall cross-sectional area of the corpus callosum, indicating differences in size of fiber tracts in the cortex.

Neuron Density and Responses of Different Neocortical Cell Populations

Neuron density is lower in rats housed in EC than those housed in IC, again showing regional differences in magnitude of effect (Diamond, et al., 1964). Beaulieu and Colonnier have recently replicated this finding with domestic cats (1985). Ferchmin and Eterovic (1986) have provided striking evidence that cellular multiplication is not a factor in EC-IC effects, by showing that inhibition of putrescine synthesis (resulting in inhibition of cellular proliferation) does not reduce EC-IC effect on cortical weights. The change in neuron density mainly reflects increased ramification of dendrites, as will be seen below. This change is accompanied by an increase in glial cell count in EC, resulting in a reduction of neuron/glia ratio (Diamond, Law, Rhodes, Lindner, Rosenzweig, Krech & Bennett, 1966). This alteration in glia cell count is primarily the result of EC-induced increases in oligodendroctyes (compared to either SC or IC), although there is some evidence of a smaller magnitude and later-occurring increases in astrocytes (Szeligo, 1977; Szeligo and Leblond, 1977). This probably indicates that enriched-experience rats have higher levels of activity in the cerebral cortex, since many functions of glial cells are involved in providing metabolic support for neuronal activity. This interpretation is supported by the difference in capillary size reported by Diamond, Krech, and Rosenzweig (1964), and the increase in relative volume of tissue cross-section samples occupied by capillaries reported by Sirevaag and Greenough (1986).

Neuronal Structure in the Neocortex

The differences observed in gross anatomical measures between animals from enriched and impoverished environments, in the absence of differences in neuronal numbers, implies the existence of differences in the characteristics of individual neurons. Inspection of individual cells provides evidence of the pervasive nature of the changes brought about in the nervous system by experience. Diamond, Lindner, and Raymond (1967) found that the cross-sectional area of neuronal nuclei was significantly larger in EC than IC, as was area of the perikaryon, a finding replicated in cats by Beaulieu and Colonnier (1985).

Using a type of analysis first described by Sholl (1956), Holloway (1966), in a preliminary study, counted intersections between two-dimensional projections of dendrites from layer II stellate neurons in the occipital cortex and concentric rings spaced at 20 micron intervals around the cell body (a simple measure of the total size of the dendritic tree). He reported greater numbers of intersections in EC tissue among layer 2 stellate neurons in occipital cortex. Volkmar and Greenough (1972) expanded this finding by quantifying dendritic branching as well: the first bifurcation in a dendrite, as it is traced away from the cell body, creates second-order dendrites, the next creates third-order dendrites, and so on. Employing this system, they found that EC rats showed consistently more higher-order dendritic branches than their IC littermates in pyramidal neurons (layers II, IV, and V) and stellate neurons (layer IV).

The volume within the neocortex affected by individual pyramidal neurons can be estimated by establishing the outer boundaries reached by the neuron's dendrites and calculating the volume of a hypothetical cylinder defined by the endpoints of the terminal, oblique, and basal dendrites. Although there have been reports that the total volume is larger in EC than IC rats (Globus, Rosenzweig, Bennett & Diamond, 1973), other researchers have reported that the increased dendritic arborization in enriched rats occurs within an equivalent volume as that occupied by the dendrites of impoverished animals (Greenough and Volkmar, 1973).

Greenough and Volkmar (1973) established that the EC-IC difference in dendritic branching becomes more pronounced as order of dendritic branching increases. Greenough, Volkmar, and Juraska (1973) found regional differences in environmental effects on dendritic branching in weanling rats comparable to the regional differences in environmental effects on brain weight and thickness: EC-IC differences were found on branching in the temporal but not frontal regions of the cortex. Although Uylings, Kuypers, Diamond, and Veltman (1978) report increased dendritic branching in occipital cortex in young adult EC rats, these investigators also report different results from those of Greenough, et al. (1973) for the frontal cortex samples. As measured by cortical thickness, Uylings, et al. (1978) found significant EC-IC differences in thickness for frontal regions (analysis of dendritic branching in this area was not reported). The apparent contradiction in these results has not been addressed.

Kopcik, Juraska, and Washburne (1986) report that EC rats have significantly lower density of unmyelinated axons in the corpus callosum, although the larger magnitude of the increase in size of this structure reported by Juraska and Meyer (1985) indicates that EC rats have a greater total number of unmyelinated axons. This increase in number of unmyelinated axons in EC rats was found without sex differences. In this study, however, female rats, but not male rats, also showed evidence of significant increases in the number of myelinated axons with enriched experience. (See Chapter 4 for a discussion of sex differences in environmental responses.)

All cell types within an area, however, are not equally plastic. An examination of several cell populations, in samples from occipital cortex of rats differentially housed in adulthood, showed environmentally induced variations in some types of cells but not in other cell types (Juraska, Greenough, Elliot, Mack & Berkowitz, 1980). For example, significant differences were found in dendritic length for pyramidal cells from cortical layer III and stellate cells from layer IV, but not for pyramidal cells from layer V. The functional significance of the cell-specific responses remains to be demonstrated.

Synaptic Anatomy in the Neocortex

It has long been presumed that changes in neural anatomy important for memory would involve alterations in the structure or efficiency of synapses, but only since approximately 1970 have neuroanatomical techniques advanced sufficiently to make the search for relationships between synaptic connections and behavioral variables methodologically practical. The general characteristics of neuroanatomical responses to enriched and impoverished environments have been fairly well described for several years, but important advances in describing neuroanatomical responsiveness to environmental manipulation in recent years have come in measures of neuronal microstructure, particularly synaptic anatomy.

Detailed examination of dendritic anatomy reveals that there are multiple characteristics of neuronal microstructure affected by enrichment or impoverishment. Enriched-experience rats have increased relative density of dendritic spines (measured per unit length of dendrites) compared to IC littermates (Globus, et al., 1973). Since dendritic spines are locations for synaptic connections, this increase in spine density is an indirect indication of increased numbers of synapses in EC rats. The difference in spine counts is most pronounced on basal dendrites, which receive input from nearby cortical cells. The next largest difference was observed in terminal and oblique segments of apical dendrites, which are relatively distal to the cell body and receive input from relatively distant cells. No consistent pattern of differences was observed in the segments nearer the cell body.

Measures of the anatomical components of intercellular connections have been studied more directly, and have also been shown to be affected by the complexity of the environmental situation. Diamond, Lindner, Johnson, Bennett, and Rosenzweig (1975) measured postsynaptic thickenings as an index of synaptic location and density, and reported fewer synapses per field of view in microscopic examination in EC rats than in IC, but this form of measurement does not correct for the decrease in neuronal density that necessarily accompany increases in the size of a brain region without a corresponding increase in neuronal number. Walsh and Cummins (1976), in contrast, reported significantly greater occipital cortex synaptic density in their EC rats. Bhide and Bedi (1984b) gathered electron micrographic evidence indicating a higher synapse-to-neuron ratio in EC rats. Estimating synaptic numbers by measuring presynaptic variables rather than postsynaptic features, Sirevaag and Greenough (1986) found that EC rats exceeded IC in numerical density of axonal boutons. It is possible that the criteria used in these studies for selection of the synapses to be measured were sufficiently different that the lack of agreement between them may have been due to the measurement of different types of synapses. It has been reported, for example, in recent studies of the visual areas of the cat neocortex, that environmental enrichment leads to a relative decrease in the overall numerical density of synapses, but that this decrease is mainly due to large decreases in the density of a particular subpopulation of synapses (Colonnier and Beaulieu, 1985). While asymmetrical synapses associated with round synaptic vesicles were unaffected by environment, symmetrical synaptic connections associated with flat vesicles were decreased by 45% in this experiment. The decrease in numerical density of synapses was accompanied by increase in the diameter of the associated boutons in enriched cats (Beaulieu and Colonnier, 1986).

Both Diamond, et al. (1975) and Walsh and Cummins (1976) found that synaptic junction cross sections were longer in EC rats than in IC littermates. Turner and Greenough report similar findings using stereological calculations for the three-dimensional reconstruction of synapses and a correction for group differences in neuronal size made possible by making the assumption that each synapse assumes a regular geometric shape; in this case synapses were assumed to be disk-shaped (1983, 1985). In this study, the number of synapses per neuron in layers I - IV of occipital cortex was found to be significantly higher in EC than IC rats.

Total area of synaptic contact and average synaptic size (indicated by length of synaptic thickening) have been reported to be greater in the occipital cortex of EC than IC rats (Mollgaard, Diamond, Bennett, Rosenzweig & Lindner, 1971). The quantitative estimates in that paper were later retracted by Diamond, et al. (1975), but the basic finding that synaptic size is plastic was maintained. West and Greenough (1972) also reported longer postsynaptic thickenings in occipital cortex for EC than for IC rats. Further study (Greenough, West, and DeVoogd, 1978) revealed a significantly higher proportion of occipital cortical synapses with perforations in the subsynaptic plate in EC than in IC rats. Sirevaag and Greenough (1985a) have reported that cross-sectional synaptic length is significantly increased in layer IV of EC occipital cortex, but when subsynaptic plate perforations (SSPP's) are considered, reducing the synaptic length by the total length of SSPP's for that synapse, the total postsynaptic thickening lengths were not different between the groups. In addition, Sirevaag and Greenough (1985) describe a population of very large synapses in layer IV of EC rats that is not present in their IC littermates, and demonstrate that the maximum synaptic length (at cross section) is significantly greater in EC than IC rats. Although there were no overall group differences in the width of the synaptic cleft, there were differences in synaptic cleft width when areas of SSPP were compared to nonperforated areas (clefts were narrower at SSPP's than at points of postsynaptic thickening); since EC animals had higher numbers of SSPP's, any effects of environmental condition on synaptic clefts would operate through environmental effects on subsynaptic plate perforations.

Studies of synaptic contact curvature have also shown this characteristic to be altered by experience in enriched and impoverished environments (Wesa, Chang, Greenough, and West, 1982): EC rats showed greater presynaptic concavity than IC littermates. This structural feature has been proposed to indicate greater synaptic efficiency (Dyson and Jones, 1980), and has the physical effect of increasing the area of a synapse without a corresponding increase in the linear proportion of cell membrane (axon, dendrite, or spine) occupied by that synapse. In other investigations, synaptic curvature has also been shown to be altered by training (Wenzel, et al., 1977a, 1977b). This evidence does, however, point to a potential problem with the interpretation of the otherwise interesting results of Turner and Greenough (1983, 1985): If synapses in EC and IC rats have different curvatures, then it may not be warranted to assume for both groups a common three dimensional disk best describing synaptic shape that was necessary in the stereological calculations employed in that study.

Greenough, Hwang, and Gorman (1985) report higher levels of polyribosomal aggregations in the postsynaptic region housed for 30 days in EC. Because location of such aggregates is an indication of synapse formation, this finding suggests that neural activity concomitant with responses to environmental enrichment may actively induce synapse formation.

The experientially-related alterations in neural and synaptic structure have been confirmed and extended by recent advances in the study of synaptic changes associated with habituation, sensitization, and associative learning in the marine invertebrates *Aplysia californica* (cf. Hawkins and Kandel, 1984), *Hermissenda crassicornis* (cf. Alkon, 1985), and other invertebrate model systems. In addition, these intensive programs of investigation have shown changes in cells specifically implicated in learning. This suggests the possibility that the neuroanatomical and synaptic alterations associated with alterations of experience may provide important clues into the neuroanatomical basis of memory in invertebrate as well as mammalian nervous systems.

Anatomical Changes Outside the Neocortex

Although the majority of experimental investigations, particularly those dealing with microstructural differences in neuroanatomy, have dealt with neocortical effects, there is evidence that other structures in the brain are also modified by experience.

Hippocampus

Experiential influences on the hippocampus are of particular interest because of recent hypotheses about the role of the hippocampus in memory, whether as a site for storage and processing of spatial information (cf. O'Keefe and Nadel, 1978), or as a device for memory indexing (Teyler and DiScenna, 1986). Walsh, Budtz-Olsen, Penny, & Cummins (1969) reported increases in hippocampal thickness as a result of enriched experience, but other investigators found only weak evidence for changes in the weight of the hippocampus (Rosenzweig and Bennett, 1978) after differential experience. Although Jones and Smith (1980) likewise found little evidence for changes thickness in several regions of the hippocampus, their discussion section makes it plain that neither was the "EC" in their laboratory particularly environmentally enriched, nor was the "IC" impoverished (other than socially). The combined reduction in the environmental difference between EC and IC makes incorporation of these results with those of other studies difficult. Katz and Davies (1983) reported that they did not find reliably significant differences in the thickness of hippocampal cross-sections at any particular location, but that the total area of the hippocampal cross-sections was significantly increased in EC rats. At a more detailed level, enriched-experience rats have been reported to have significantly higher numbers of granule cells in the dentate gyrus than impoverished-experience littermates (Susser and Wallace, 1982). Walsh and Cummins (1979) did not find significant EC-IC differences in the size of hippocampal neuronal nuclei, although they did report that nuclear size in the granular layer was more variable among IC than EC rats.

Although Fiala, Joyce, and Greenough (1978) found increases in some aspects of dendritic branching and overall size of the dendritic field in the dentate gyrus, this effect was found only for juvenile, but not adult, rats. (The possibility of critical periods in environmental effects will be discussed in Chapter 5.) Altschuler (1979) found that ECT rats had increased synaptic density in area CA3 when compared with either a motor activity control group or "standard" conditions (it is not clear whether this is equivalent to IC or animals were housed in social conditions, or SC).

Cerebellum

Structural plasticity of the cerebellum has been demonstrated in the Japanese macaque (*Macaca fascicularis*) as a function of differential environments (Floeter and Greenough, 1978, 1979). Monkeys reared under colony conditions (large rooms with social stimulation as well as climbable and manipulable objects) until 8 months of age showed significantly larger Purkinje cell bodies in the nodulus and uvula of the cerebellum than isolation-reared monkeys or monkeys who had limited social experience but were not reared in the colony. Colony reared animals also had more dendritic material (as assessed by an elaboration of the concentric ring analysis first described by Sholl, 1956) than socially-reared monkeys. This was found in the Purkinje cells of the nodulus, which develops relatively early (Altman, 1969), and in the paraflocculus, which develops later. Granule cell dendritic fields in these same areas, however, did not differ, nor did Purkinje cell dendritic fields in the flocculus.

Although the weight of brain regions outside the forebrain (specifically the neocortex and hippocampus) have not been shown to exhibit substantial plasticity (Bennett, et al., 1964a), in recent studies by Greenough, McDonald, Parnisari, and Camel (1986) some aspects of microstructural anatomy in the cerebellum do show plastic response to environmental alteration. In rats 24-26 months of age at the start of the experiment, EC subjects showed more spiny branchlets (but not main branches) of cerebellar Purkinje cells than rats housed socially in small cages without stimulus objects. Comparisons of EC with isolated animals given opportunity to exercise (Black, Parnisari, Eichbaum, and Greenough, 1986) revealed that EC subjects had more material in mid-region spiny branchlets and less at the ends. These results, as well as reports by Thompson and co-workers concerning the cerebellar role in some types of Pavlovian conditioning (cf. Thompson, 1986) suggest that cerebellar plasticity may have been improperly neglected as a topic for investigation.

Other Non-neocortical Structures

Although the neocortex and associated structures are reliably different in rats housed in EC and IC conditions, the pattern of EC-IC differences in gross anatomical measures of the subcortical structures of the brain is less clear. In the standardized dissection procedure used in the Berkeley laboratories, the tissue removed as part of the four cortical samples includes (in the ventral cortex sample) adjacent tissues such as corpus callosum, hippocampus, and amygdala. Bennett, et al. (1964a) presented evidence from multiple experiments that the remaining brain parts (including the brainstem, midbrain, cerebellum, and thalamus) actually show a slight decrement in weight in EC as compared to IC rats. This difference was reliable across experiments, but of sufficiently small magnitude that it was statistically significant only when results from seven experiments were pooled. Quay, Bennett, Rosenzweig, and Krech (1969) found that, when adjusted for group differences in body weight, the pineal organ shows no EC-IC differences.

In view of the small magnitude of EC-IC subcortical differences in tissue weight, and the correlation between total body weight and brain size, a ratio of cortical to subcortical weight can be used to provide an indication of cerebral effects of experience that is relatively independent of group differences in body weight. This yields a measure more easily compared across experiments, in cases where the interpretation of absolute values for brain measures would be complicated by body weight differences. Use of this measure provides evidence of EC-IC cerebral differences that is quite stable across multiple experiments, with EC exceeding IC by 5% (Rosenzweig, Bennett, and Diamond, 1972c).

Neurochemical Changes in Response to Differential Experience

As is the case with neural structure, many aspects of neurochemistry have been examined for changes subsequent to differential experience. These include total content and concentration of the nucleic acids RNA and DNA, brain protein content, and measures indicating the concentrations and turnover rates of specific neurotransmitters (including the cholinergic system, as indicated by the enzymes cholinesterase and acetylcholinesterase) and neurotransmitter receptors. The results of these investigations will be discussed in the sections below.

Nucleic Acids

Bennett (1976) reported lower DNA density in cortical tissue for EC rats, measured as DNA per milligram of tissue. This corresponds to the findings of Diamond, et al. (1964, 1966), describing unchanged total cell counts in EC rats, and can be explained by reference to the increase in cortical weight without a corresponding increase in cell numbers: Since DNA is present in a fixed amount per cell, and cell numbers do not change with enriched experience, an increase in cortical weight necessarily leads to a decrease in DNA density. Ferchmin, Eterovic, and Caputto (1970) reported that EC rats had more cortical RNA than IC littermates. The ratio of RNA/DNA has yielded consistent differences of between five and six percent between EC and IC animals, approximately equal to the change in overall cortical weight. As the primary function of RNA is manufacture of proteins, this increase in RNA per cell is indicative of a heightened activity of cells in the EC subjects.

After 30 days of differential environmental treatments, brain RNA showed more diversity of sequence in EC than IC animals, while no differences in sequence diversity were observed in liver RNA from the same animals (Grouse, Schrier, Bennett, Rosenzweig, and Nelson, 1978). Subcortical measures of RNA and DNA show little, if any, change with variations in environmental complexity; the increases in metabolic activity shown by changes in nucleic acids are limited to the cerebral cortex. Taken together, these data suggest that not only are the cells in the cerebral cortex of EC rats more metabolically active in a general sense than are their IC littermates, but they are also producing proteins in greater quantity and diversity.

Protein

Bennett, Diamond, Krech, and Rosenzweig (1969) reported that brain protein content varied directly with wet tissue weight (the measure typically employed as an index of EC-IC effects in studies concerning gross cortical anatomy), and that environmentally induced differences in brain protein are of an identical magnitude as differences in tissue weight.

In addition to measures of overall protein content, particular structural proteins are also altered. Jorgensen and Meier (1979), in addition to replicating that differences in overall weight and differences in protein content were of identical magnitude, found significant increases in colchicine binding, indicative of higher levels of tubulin dimer (the structural components of axonal microtubules), in EC rats. Similar findings were also obtained with two tubulin-related antigens, T-antigen (which binds to tubulin-oligomer) and NT2 (which binds to microtubular associated protein). The most intriguing finding here, however, is that two of the three measures (colchicine and T-antigen, binding to tubulin dimer and oligomer) indicate that the increases in this study in tubulin-related proteins were significantly larger than the general increase in tissue protein. This would seem to indicate that, in rats, the EC-IC difference in microtubules is more pronounced than the changes in many other anatomical and chemical measures. The functional significance, if any, of this particular reaction to differential environments is not known at present.

Recent evidence indicates that cortical putrescine is lower in EC than IC rats (Eterovic and Ferchmin, 1986), which the authors suggest (Ferchmin and Eterovic, 1987) may indicate an increase in cell differentiation in EC subjects.

Cordoba, Yusta, and Munoz-Blanco (1984) examined response to environmental manipulation of amino acid concentration in mice, in an interesting study, but one for which confidence in the findings is limited by small sample sizes (n=6 per group) and the lack of an attempt to replicate these findings. They report that EC mice differ from both SC and IC mice in the amino acid aspartate in the spinal cord, of glutamate in the colliculi and cerebral cortex, and glycine in the pons-medulla, colliculi, and cerebral cortex. Other amino acids (threonine, serine, alanine, isoleucine, and leucine) were not affected by environmental treatment in any brain region studied.

Uphouse (1978) has described increases in the capacity of cortical chromatin to support RNA synthesis in rats after 30 days housing in an enriched environment. Later reports, however (Uphouse and Tedeschi, 1979) showed that this difference disappears in rats differentially housed for 60 days. The confidence we can place in these results is, however, impaired by the small sample sizes employed in these studies.

In view of the evidence that synthesis of protein, in which RNA plays a crucial role, is centrally involved in the formation of long-term memory (cf. Rosenzweig and Bennett, 1984), further work on this problem could prove highly informative. The possibility of direct measurement of a transient change in a significant metabolic variable offers the potential for exciting advances in our understanding of the physiological mechanisms that form the basis of the brain's reaction to experience.

Cholinergic System

The cholinergic system has long been implicated in multiple behavioral systems, including learning and memory. The first result indicating that significant characteristics of the nervous system might be influenced by experience (Krech, Rosenzweig, and Bennett, 1960) was essentially a serendipitous outgrowth of studies based upon the proposition that individual differences in cholinergic function might be related to individual differences in learning ability (Rosenzweig, Krech, and Bennett, 1960). In these studies, rats performed tasks for assessment of learning ability, and were subsequently sacrificed to allow measurement of cholinergic function. It then appeared that the rats trained on more difficult tasks or trained more extensively had higher values of the general enzyme cholinesterase. The enriched and impoverished environments were then employed to reduce the investment of time on the part of laboratory personnel required for formal training.

When compared to IC littermates, EC rats showed increases in the total activity of the general enzyme cholinesterase (ChE) in the cortex (Rosenzweig, Krech, Bennett & Diamond, 1962). When corrected for changes in tissue weight, the ChE levels of EC rats were still significantly higher in EC than IC rats (Rosenzweig, Bennett, and Diamond, 1972c). In the rest of the brain, no consistent pattern of differences has emerged over multiple studies.

Although rats housed in EC were initially reported to have greater total activity of the specific enzyme acetylcholinesterase (AChE) in both cortex and subcortex (Bennett, Diamond, Krech & Rosenzweig, 1964), later studies present a more complex picture. Rosenzweig, et al. (1972c) report a trend (over 16 experiments conducted between 1963 and 1969) that did not reach statistical significance toward slightly higher total brain AChE levels in EC rats. Activity of AChE per unit weight is slightly lower in the cortex and slightly higher in subcortex for EC rats, and a relatively stable EC-IC difference is found in the cortical/subcortical ratio of relative AChE activity (Rosenzweig, et al., 1972b). In addition, the difference in AChE is significantly larger in comparisons between rats housed in seminatural environments and littermates from IC than in EC vs. IC comparisons (Rosenzweig and Bennett, 1978).

The differences in environmental effects on AChE and ChE were unexpected, and the relatively high concentration of ChE in glia provided the suggestion that this might indicate a relative increase in glial cells in EC animals, a suggestion confirmed by Diamond, et al. (1966) as reported above.

Other Neurotransmitters and Neurotransmitter Receptors

Many enzymes and neurotransmitters have been assayed, with mixed findings. Pryor (1964) found no EC-IC differences in brain hexokinase, nor were cortical serotonin concentrations found to be different. Although Geller, Yuwiler and Zolman (1965) examined concentrations of serotonin, 5-HTP decarboxylase and dopamine and found no differences in whole brain analyses, they did find increases in the norepinephrine content of the whole brain in IC rats. Their procedure of weaning rat pups at 19 days of age may, however, have confounded these results by stressing the pups. Riege and Morimoto (1970) found increased norepinephrine in the cortex of EC rats along with a decrease in the hypothalamus-caudate complex, producing a whole brain net change of approximately zero. Dopamine levels showed a similar pattern of change, while serotonin was slightly (but significantly) reduced in a cortex-only sample.

There is good evidence, however, that monoamine neurotransmitter tissue concentrations are substantially altered quite rapidly postmortem by continued enzymatic activity, raising some doubt as to the trustworthiness of findings from studies employing techniques that include delays between sacrifice and procedures that inactivate enzymatic degradation of existing transmitters. One method of accomplishing immediate interruption of relevant chemical activity is the use of whole-brain microwave irradiation for sacrifice. This procedure raises whole-brain temperature sufficiently to denature the relevant enzymes in less than one second. One of the authors of this monograph (MJR) and the neurochemist C. LeRoy Blank (of the University of Oklahoma) have initiated investigations to examine the response of the monoamine neurotransmitter systems to environmental manipulation by using this technique, followed by liquid chromatography with electrochemical detection for direct determination of catecholamine and indoleamine transmitters, precursors, and metabolites. Preliminary evidence (Renner, Blank, Freeman, and Lin, 1986) appears to confirm previous findings that serotonin concentration is unaltered by differential experience, but suggests that serotonin turnover rate may be increased in hippocampus in IC rats. Cortical norepinephrine concentrations were not significantly different between EC and IC, but dopamine in the occipital cortex was significantly increased in the EC group. These investigations are continuing.

Por, Bennett, and Bondy (1982) examined the binding sites of several neurotransmitters (dopamine/serotonin, alpha-adrenergic, beta-adrenergic, muscarinic cholinergic, and GABA, as well as binding sites for benzodiazepines) for enriched and impoverished rats, and found no significant differences between the groups for any of these receptor types.

Physiological Changes in Response to Differential Experience

In the face of the diverse range of alterations in the structural components and chemical makeup of the central nervous system, it would be surprising if there were no concomitant alterations in some aspects of function as well. Although response of CNS function to experiential manipulation has not been studied as thoroughly as have the structural and chemical responses, there is some evidence that the physiology of the brain responds to enriched and impoverished environments, and further research along these lines would undoubtedly be rewarding.

Sleep

The relationship between learning and memory processes and the organization of sleep and waking, including the balance between slow-wave sleep (SWS) and rapid eye movement sleep (REM), has long attracted considerable attention among psychobiologists (see McGrath and Cohen, 1978, for a review of this area). The obviously greater exposure to environmental change in the enriched condition has prompted several investigators to examine the sleep patterns of differentially housed rodents, both rats and mice. Tagney (1973) reported that EC rats exhibit a significantly higher percentage of time spent asleep than IC, for both phases; proportion of total sleep time spent in either SWS or REM sleep did not differentiate the groups. Tagney's findings were replicated by Kiyono, Seo, and Shibagaki (1981). Mirmiran, Van den Dungen, and Uylings (1982) report that this pattern is evident by the third week in differential conditions, and the magnitude of the EC-IC difference increases with duration of differential housing. Mirmiran, et al. (1982) found that socially housed rats (SC) did not differ on these measures from IC. Gutwein and Fishbein (1980a; 1980b), using mice, replicated the EC-IC difference in REM sleep, but did not find the SWS differences reported in the studies conducted with rat subjects, described above. There are no obvious methodological differences that provide easy explanations for this discrepancy; methodologies in these studies were quite similar (all studies employed polygraphic measures taken from implanted electrodes). Perhaps the difference in SWS must be attributed to a species difference between mice and rats, but whereas the effects in rats have been reported by several laboratories, those in mice have so far been reported by only a single laboratory. The topic of effects of differential experience on different phases of sleep is clearly a feasible area of investigation and one that could reward further effort.

Neurophysiology and Electrophysiology

After only a single electrophysiological investigation on differentially housed animals in the 1960s, several reports on this topic have appeared in the 1980s. Edwards, Barry, and Wyspianski (1969) reported a decrease in the latency of a visually evoked cortical potential in rats following enrichment, relative both to IC littermates and to pre-differential housing values. Leah, Allardyce, and Cummins (1985) examined cortical potentials evoked by somesthetic stimulation (threshold-intensity electrical stimulation of a forepaw). They found habituation in evoked potentials when this stimulation was presented repeatedly to EC rats, but no habituation in IC rats. When an identical procedure was carried out after an intervening delay of one hour, however, both groups showed a decrease in amplitude of response with repeated stimulation. This physiological finding matches the behavioral pattern that EC are quicker than IC subjects to alter their behavior in response to environmental feedback, as will be discussed in the next chapter.

Evidence for enrichment-induced enhancement of synaptic function has been provided by Sharp, McNaughton, and Barnes (1983, 1984) that the first several days' experience in an enriched environment elicits changes in synaptic efficiency in the perforant path-dentate granule cell synaptic connection of the hippocampus (these findings have been replicated by Green and Greenough, 1986). The characteristics of these changes were described as being highly similar to those of the long-term enhancement found after electrical stimulation of that region. There is, of course, widespread current speculation concerning the possible role of long-term enhancement (or long-term potentiation) as a mechanism for long-term memory storage.

Cerebral Metabolism

There is considerable evidence, albeit mostly indirect, that the metabolism of the cerebral cortex is significantly different in animals with different experiential histories. As discussed above, glial cells perform multiple roles in support of cerebral neural activity, and they are found in higher numbers in EC than IC rats (Diamond, et al., 1966). Likewise, measures of capillaries indicate the potential for increased blood flow in ECs (Diamond, et al., 1964; Sirevaag and Greenough, 1986), and increases in cortical RNA (Bennett, 1976) imply increases in protein synthesis in EC cortex.

A more direct approach to questions of cerebral metabolism has been reported in an investigation of subjects' responses to administration of anesthetic convulsant drugs by Juraska, Greenough, and Conlee (1983). These investigators report faster response to anesthetics in EC rats and a lowered seizure susceptibility in IC rats under stroboscopic lighting (but not in conditions of dim illumination), indicating decreased nervous system excitability in EC rats. Greenough, Yuwiler, and Dollinger (1973) found dose-dependent EC-IC differences in Lashley III maze performance following eserine injections, even thought there was no evidence of group differences of eserine toxicity or cholinergic inhibition. When combined with the implication of increased cerebral blood flow, these results suggest that the net EC-induced change in cerebral metabolism is one of increased capacity for adaptive response to challenge.

Chapter Summary

Research reports of the 1960s demonstrated that differential experience produces measurable changes in neurochemistry and anatomy of the brain, principally in the neocortex. Many other investigators have since entered this field of research; their reports have corroborated the earlier findings and have extended them in a number of directions. For example, both the Berkeley group and other investigators have extended the gross neuroanatomical studies to measurements of dendritic branching and synaptic morphology. Work on the neocortex has been extended to the hippocampus and cerebellum, regions that are implicated in current hypotheses of memory formation. Electrophysiological changes related to differential experience have been reported in the 1980s with regard to both sleep patterns and to synaptic potentials.

We have tabulated the research reports cited in this chapter on effects of differential experience on neurobiological measures for each 5-year interval beginning with 1960-64. The tabulation shows a rise in the number of reports after the first period, and then a relatively steady rate of productivity thereafter. It appears that the study of effects of differential experience on the nervous system continues to be fruitful.

Chapter 3:
Behavioral Effects of Differential Experience

Just as manipulation of the complexity of the stimulus environment leads to changes in the brain, experimental manipulation of the stimulus world has a measurable impact on behavior; indeed, this issue was studied prior to the search for neural correlates of differential experience. As early as 1947, Hebb reported behavioral differences between rats reared in a complex environment and rats reared in an impoverished environment. Since that time, the majority of investigations of behavior in differentially housed animals have focussed on direct measures of learning and memory; few studies have addressed other facets of behavior. While an impressive catalog can be assembled of our knowledge of anatomical and neurochemical differences between the brains of animals from enriched and impoverished environments, comparatively little is known about just what these differences are good for in the animals' world, and even less is known of the nature of the relationship between cerebral and behavioral alterations induced by differential environments.

The fundamental assumption that organizes all psychobiological endeavors is that behavioral events are somehow governed by biological events, most frequently neurobiological ones, and most directly those occurring within the central nervous system. All too often, however, the search for the biology underlying behavior becomes investigation into biology for its own sake, with behavior somehow lost in the shuffle, or so far removed that the connection is not apparent. As Corinne and S. J. Hutt put it: "To correlate behavioral measurements of the crudity of better-worse, more-less with physiological variables measured to two decimal placements in micrograms is, to say the least, faintly ridiculous."

In studying the effects of differential environments on brain and behavior, the obvious implication from findings that animals with an enriched history exceed their impoverished counterparts in many brain measures is that they will also be behaviorally superior. This inference often goes unstated and even unexamined. The ever-increasing detail available in our understanding of neural effects of experience in differential environments demands a parallel expansion of the understanding of behavioral consequences of differential housing and examination of the relationship between brain and behavioral changes. Although animals housed in complex environments *are* behaviorally different from those housed in impoverished environments, the linkages between brain changes and alterations in behavior are not obvious.

While there have been many studies, both those reporting an absence of differences between performance of EC and IC subjects and those reporting behavioral differences indicating some type of superior performance by EC animals, interpretation of these studies is rendered difficult by the variability in definitions of what constitutes environmental enrichment. This problem is particularly pronounced in studies of the behavioral effects of enriched and impoverished experience, which come from a variety of sources, including traditional learning paradigms as well as directly from EC-IC research. There is less of a problem in defining enriched environmental conditions in biological studies of this type, because nearly all of these trace direct historical and intellectual antecedents from the early studies of the effects of differential environments on the brain performed in the Berkeley laboratories.

This chapter examines behavioral differences between animals with enriched and impoverished experience. We will address, in order, possible differences in social behavior, measures of problem-solving effectiveness, and the organization of spontaneous behavior.

Social Behavior

An important part of the world of most species of rodents is the realm of social interaction. *Rattus norvegicus*, especially, is highly social, and it might be expected that some aspects of social interaction could be affected by manipulations of the environment that produce measurable alterations in the nervous system. In spite of the plausibility of this hypothesis, however, we know of no investigation wherein the social behaviors of enriched and isolated subjects were compared directly to each other. The examination of the social behaviors of differentially housed animals must be parcelled into examinations of alterations produced by isolation and those produced by enrichment, comparing each group to socially-housed subjects.

The literature concerning the effects of impoverished experience on later behavior is voluminous, and we will not attempt a complete review here. In summary, isolated rats have a higher tendency towards aggressive behavior (a wealth of literature exists here; see, for example, Baenninger, 1967), and isolated rats are less skilled in providing aggression-inhibiting cues to other rats (Luciano and Lore, 1975). These effects may, however, have a critical period, developing only when isolation occurs before age 50 days, and may occur only in species that engage in high levels of social play (Einon, Humphreys, Chivers, Field, and Naylor, 1981); neither limitation applies to cerebral effects or to alterations observable in other aspects of behavior, so there is reason to doubt that these isolation effects are due to the same factors as are other EC-IC effects.

In an extensive review of the effects of different types of housing on behavior in rats, Dalrymple-Alford, Benton, and Brain (1983) were unable to find reports contrasting the social behaviors of group-housed animals with those housed in enriched conditions. In direct comparisons of social interaction patterns in separate groups of adult and juvenile rats, housed in grouped or enriched conditions, Renner and Rosenzweig (1986a) found that adults showed lower activity levels than juveniles, but found no evidence of reliable differences in social interaction between the enriched and impoverished groups in either adults or juveniles. Although little evidence has emerged to date of any such environmentally induced effects, it remains possible that some specific aspects of social interaction (e.g., relative skill in providing clear signals in social communication with conspecifics) are altered by differential environments. In view of the importance of social interaction in the everyday existence of this species, further research into possible environmental influences on social interaction is clearly warranted.

Learning

Numerous investigators have reported results of studies designed to reveal differences in behavioral or information processing abilities in animals from enriched and impoverished environments. These studies have employed problems ranging across a broad spectrum, and present a clear pattern of results: In problem-solving tasks, the more complex the task, the greater the likelihood that EC-IC differences will be found. Further results have led to a variety of interpretations, as we will see, and research in this area is being pursued actively.

Relatively simple tasks do not yield consistent EC-IC differences. Domjan, Schorr, and Best (1977) did not find group differences in taste-aversion learning, and Van Woerden (1986) found EC and IC equivalent on rate of habituation to repeatedly presented acoustic startle stimuli. Caul, Freeman, and Buchanan (1974) did not find EC rats different from IC on acquisition of conditioned heart-rate suppression. Although some investigators have reported EC superiority on acquisition of visual discrimination (Edwards, Barry, and Wyspianski, 1969; Brown and King, 1971; Bernstein, 1973), others found no such difference (Bingham and Griffiths, 1952; Woods, Ruckelshaus, and Bowling, 1960; Krech, Rosenzweig, and Bennett, 1962; Gill, Reid, and Porter, 1966; Sjoden, 1976). Likewise, active avoidance tasks yield mixed results: both significant EC superiority (Ray and Hochhauser, 1969) and the absence of differences (Doty, 1972; Freeman and Ray, 1972) have been reported. Van Woerden (1986) offers evidence that the relative novelty of the cue stimulus affects EC and IC differently: in subjects for which the cue stimulus was novel, there were no EC-IC differences on a discrimination task, but for subjects pre-exposed to the cue, EC rats outperformed those from IC.

Ough, Beatty, and Khalili (1972) found identical rates of acquisition of a barpressing task, but reported that EC rats are superior at response inhibition, as measured by schedules of reinforcement that differentially reinforce low rates of responding (i.e., DRL schedules). Lore (1969) also reported that EC rats are more successful than their IC counterparts at passively avoiding a candle flame; Freeman and Ray (1972) also found EC rats superior on other forms of passive avoidance tasks, but there have been contrary reports. Davenport (1976) reported that, although enriched experience aided in behavioral recovery from experimental hypothyroidism (discussed in Chapter 6), non-treated EC and IC did not differ on passive avoidance tasks.

The most consistent finding across studies, although it is not universal, is that of superior performance by EC animals in complex problem-solving tasks. EC rats do exhibit an advantage on tasks involving reversal of visual discriminations learned previously (Krech, et al., 1962; Doty, 1972; Bennett, et al., 1970; the lone nonsupporting result comes from one of several groups within Bennett, et al., 1970). EC rats also show advantages on other forms of response flexibility (Nyman, 1967, using alternation learning). Van Woerden (1986) has demonstrated differences in spatial reversal problems and has shown further that the EC-IC difference increases if irrelevant cues are added to the situation. Morgan (1973) reported results for varied learning tasks with rats from enriched and impoverished conditions, for example, removing an obstacle to enter a food compartment. EC and IC did not differ in acquisition rate for these tasks, but the enriched rats were superior in a transfer test where they were required to remove the obstacle in a different way than they had used previously.

Many studies comparing learning in enriched-housed and impoverished-housed animals have used spatial problem-solving tasks, most commonly the Hebb-Williams mazes (Hebb and Williams, 1946; typically used according to the protocol described by Rabinovitch and Rosvold, 1951), which was originally described as an intelligence test for animals. Most studies with this series of maze problems have found enhanced learning in enriched-experience rats (e.g., Hymovitch, 1952; Forgays and Forgays, 1952; Cooper and Zubeck, 1958; Woods, 1959; Denenberg and Morton, 1962b; Brown, 1968; Bennett, Rosenzweig and Diamond, 1970; Smith, 1972), and cats (Wilson, Warren, and Abbott, 1965; Cornwell and Overman, 1981). There have also been reports, albeit fewer in number, of failure to find differences in rats (Hughes, 1965; Reid, Gill and Porter, 1968). Maze performance of EC rats is more disrupted than that of ICs by rotation of the maze (Hymovitch, 1952; Forgays and Forgays, 1952; Brown, 1968); EC rats are, therefore, more able to use (or are more dependent upon) extra-maze cues in the solution of spatial problems. Luchins and Forgus (1955) reported that EC are quicker than IC to abandon a previously forced indirect path through a maze, when the conditions forcing the indirect path are removed;

this is consistent with both enhanced use of extra-maze cues by EC and enhanced ability to perform reversals.

Comparisons of rats with enriched and impoverished histories on different spatial problems yield generally similar conclusions. Studies using the Lashley III maze (Lashley, 1929) are less common than Hebb-Williams maze studies, and all reports are of superior performance by EC rats (Ray and Hochhauser, 1969; Bennett, Rosenzweig and Diamond, 1970; Riege, 1971; Freeman and Ray, 1972; Greenough, Madden and Fleischmann, 1972; West and Greenough, 1972; Bernstein, 1973; Greenough, Yuwiler, and Dollinger, 1973). Although Greenough, Wood, and Madden (1972) reported EC mice superior to IC on the Lashley III maze, Warren, et al. (1982) did not find EC-IC differences for mice on this task. In a 17-arm radial maze, EC rats learned more quickly and accurately than IC rats, as measured by number of correct choices prior to the first error, by total errors, or by total number of correct choices in the first 17 choices made (Juraska, Henderson, and Muller, 1984). In addition, these investigators reported that there were no effects of gender in environmental effects on radial maze performance.

The experientially-induced alterations that lead to differences in behavior may be relatively long-lasting: Denenberg, Woodcock, and Rosenberg (1968) report that EC-IC differences in Hebb-Williams performance are significant in female rats, even if a 300-day delay (by housing subjects from both groups in standard colony conditions) is imposed prior to the start of testing. At that time, Denenberg, et al. interpreted these results as evidence that EC-IC brain changes are permanent. Subsequently, however, it has been shown that the neurobiological alterations induced by differential experience are at least partially reversible (see Chapter 4 for discussion of this topic). A more robust explanation may be offered: The information-processing changes resulting from experience in EC survive other neural alterations which may be more specific in nature. These surviving changes would then contribute to later behavioral differences.

As proposed at the beginning of this section, it is quite apparent that there is a positive relationship between the difficulty of the problem presented to subjects with enriched and impoverished experiential histories and the probability that meaningful differences will be observed in their behaviors. This difference in environmental effects on simple and complex tasks has been repeatedly demonstrated *within* groups of subjects. We will not catalog here every instance of this phenomenon, but rather give only a few examples: Krech, et al. (1962) reported significant EC-IC differences in visual discrimination reversals, although there were no differences in acquisition of discrimination. Domjan, Schorr, and Best (1977) did not find group differences in taste-aversion learning in rats, even in the presence of differences in Hebb-Williams performance. Warren (1985) showed the performance of mice to be facilitated by enrichment on Lashley and Stone mazes, but not changed on several simple tasks (e.g., brightness discrimination and reversal, memory for location of water in an arena, spatial discrimination and reversal).

While there is a preponderance of evidence that enriched animals outperform their impoverished counterparts, attempts to interpret this behavioral difference have resulted in a variety of explanations. Woods (1959) found that EC-IC differences in Hebb-Williams performance decreased (while remaining significant) if the trial was ended at first entry to the goal box rather than with eating. (Most subsequent investigators have altered the Rabinovitch and Rosvold protocol to end each trial at the time of goal box entry.) Woods, Fiske and Ruckelshaus (1961) manipulated degree of food deprivation, and reported that they found differences in favor of EC rats in a low-drive condition, but that these differences disappeared in a high-drive condition. They interpreted this result as evidence that performance is less disrupted by the subject's curiosity about the situation in EC than IC, although this interpretation is not consistent with direct inquiries concerning the effect of differential environments on investigatory behavior (see the section below on the organization of spontaneous behavior). Van Woerden (1986) argues that the known differences in function of cholinergic brain systems create superior ability for stimulus selection in EC rats, and argues that many EC-IC performance differences are the byproducts of differences in stimulus selection.

The interpretation of EC-IC differences on tasks mediated by presentation of exteroceptive stimuli is opened to question, however, by the recent report of Rose, Love, and Dell (1986) that the relationship between the brightness of a barpress-contingent light (1 second illumination) and its effect on barpressing differs for EC and IC rats: brief presentations of light become aversive at lower intensity for EC than for IC rats. If, in fact, the same physical stimulus may carry a different significance for IC than for EC subjects, as is implied by these results, there are important implications for changes in behavioral organization brought about by environmental differences. In addition, behavioral differences discovered through the use of tasks involving punishment (e.g., passive and active avoidance) may have to be reinterpreted, as footshock of a particular intensity may be perceived as differently aversive by the two groups, and their subsequent performances could not then be clearly ascribed to differences in information processing or behavioral abilities. This interpretation is supported by the reports of Woods, et al. (1961), that EC and IC respond differently to manipulation of level of food deprivation, of Juraska, et al. (1983) that IC rats show lower convulsion threshold than EC (under metrazol, in stroboscopic lighting conditions, but not in dim steady light), and of Van Woerden (1986), that the performance of EC rats is less disrupted than that of IC by irrelevant cues in discrimination problems. Rose, et al. (1986) also report that spontaneous barpressing rates (with no barpressing contingency other than the mechanical noises associated with depressing the bar) are reliably lower in EC than IC rats.

Dell and Rose (1986) report that the acquisition slope for Hebb-Williams maze performance does not differentiate EC from

IC, and have offered the hypothesis that EC-IC differences are due to impaired asymptotic performance of IC subjects. Their analysis, however, reports only the total number of errors per trial. It has been reported several times that a very different pattern of results is seen if initial errors and repeated errors (those made previously within the same trial) are separated; EC and IC differ little on the number of initial errors, but EC make substantially fewer repeated errors. The hypothesis offered by Dell and Rose (1986) concerning inadequacy of response inhibition as an explanation of EC-IC performance differences may, in fact, be partially correct: That low ability to inhibit responses could be a contributing factor in these differences is consistent with the findings of Lore (1969), Freeman and Ray (1972), and Ough, et al. (1972). Deficiency of response inhibition cannot, however, account for all problem-solving differences between animals with enriched and impoverished histories; no explanation of the existing behavioral differences documented thus far can account for the variety of findings without including some type of cognitive difference between EC and IC subjects.

Consistent with neurobiological evidence that the primary sites of environmentally-induced anatomical plasticity are in those regions of the brain associated with the more complex (and presumably higher-level) cognitive functions (i.e., cerebral cortex, hippocampus, cerebellum), the primary evidence of environmental alterations in task-driven behavior is to be found in performance of those types of tasks requiring higher-level problem-solving skills. This would seem to suggest that EC-IC differences might be found on tasks requiring declarative learning, in the absence of differences on tasks requiring procedural learning (labelled, respectively, as memories and habits by Mishkin, Malamut, and Bachevalier, 1984). Further investigations of the specific nature of EC-IC differences in ability to learn, remember, and process information would undoubtedly be rewarding, and could in addition suggest potential applications of these findings.

Organization of Spontaneous Behavior

The broad spectrum of task-driven behavioral differences between enriched and impoverished animals leads naturally to questions concerning the nature of the animals' spontaneous behavior following this type of treatment. Study of the influence of environmental history on the behavioral predilections of experimental subjects might lead to deeper understanding of and clues concerning the nature of the environmentally-induced neurobiological changes, and might also inform behavioral scientists about questions of animal information processing. Little, however, has yet been accomplished in this direction. Unless we make the assumption that performance differences on experimenter-imposed laboratory tasks capture the bulk of the functions of animal behavior, the gaps in our knowledge of the behavioral effects of differential experience are strikingly large. Investigators have only infrequently asked the question: how do these experiences, which cause clear and broad-ranging changes in the nervous system and behaviors that are specifically problem-solving in nature, affect the way in which this animal interacts with its environment?

Much of the relatively little work that *has* been done on possible changes in behaviors not tied to arbitrary laboratory tasks in enriched and impoverished animals has been focused on the effects of differential environments on locomotion and defecation in an open field (e.g., Zimbardo and Montgomery, 1957; Denenberg and Morton, 1962a; Freeman and Ray, 1972; Smith, 1972; Studelska and Kemble, 1979). There exists, however, considerable disagreement concerning what is being measured in the open field (e.g., Denenberg, 1969). Much of the rat's behavior in an open field is best described as the rat's attempt to get out of the open field (Welker, 1957; Aulich, 1976); Suarez and Gallup (1981) have also demonstrated that exposure in the open field in the presence of an observer elicits clear-cut predator avoidance responses in chickens. Lore and Levowitz (1966) described opposite effects of enriched and impoverished environments on two different measures, both supposedly indexes of exploratory behavior. During "forced" exploration (in which the subject had no option but to remain in the arena), IC rats showed higher activity levels, whereas in "free" exploration (in which the subject could remain in a protected spot) EC rats emerged from cover sooner, indicative of a higher level of exploration. The confusion over what is measured in an open field is best demonstrated by the fact that it is described by some investigators as a measure of exploration (e.g., Freeman and Ray, 1972; Smith, 1972) and by others as a measure of emotionality (cf. Denenberg and Morton, 1962a). The open field situation may in fact elicit both emotional and exploratory behaviors (Whimbey and Denenberg, 1967, but see also Aitken, 1974, Walsh and Cummins, 1976b, 1978; Royce, 1977), but it does not do so in such a way that we can, post hoc, disentangle them. One conclusion which does seem clear is that we have been left until recently with very little interpretable information about the effects of environmental enrichment or impoverishment on behaviors not driven by specific laboratory tasks.

Renner and Rosenzweig (1986b) have examined spontaneous exploratory behavior among rats housed in EC and IC from 30 to 60 days of age, in an experimental situation specifically designed to overcome many of the characteristics of open-field tests that lead to difficulty in interpreting results. By using remote video observation, the problems inherent with the presence of the experimenter are resolved, and the use of videotape allows multiple observations of the same event; this increases both the reliability and the probable validity of the observations for drawing inferences about the character of spontaneous behavior. In this context, no EC-IC differences were found in overall level of exploratory behavior (on any of several measures) or in willingness to interact with objects (as measured by total time spent interacting or number of bouts of interaction). Significant differences were apparent in behavioral organization in interacting with objects, in that bouts of object interaction in EC rats were more complex than in IC rats. In young adult rats, both amount and organization of exploration are affected by differential experience (Renner, 1987a). These findings carry two important implications: first, that behavioral organization is altered by differential experience in a manner amenable to empirical investigation, and second, that to attempt to characterize spontaneous exploratory behavior by means of a single measurement (such as movement in an open field) is to oversimplify a complex behavioral phenomenon.

It has been demonstrated that rats from enriched conditions are superior at solving complex problems: it is widely presumed that an increase in problem-solving skill will be beneficial to the individual who possesses it. Sensitivity to real-world problems faced by the species typically used in these experiments could provide guidance as to the types of situations most appropriate for studying the question of whether there are functionally significant changes in behavior induced by differential experience. (A similar concern has been voiced in the animal learning literature [e.g., Seligman, 1970; Johnston, 1982] in calling for studies of learning motivated by ecological considerations.)

The question of the possible functional significance of environmentally-induced behavioral differences has been addressed in two studies: In the first, Roeder, Chetcuti, and Will (1980) report slight differences in patterns of survival under predation between rats with EC and IC histories. Replicable patterns of differences in numbers of EC- and IC-experienced rats surviving in the presence of a polecat (*Mustela putorius*) were shown beginning at 15 days after onset of predation, although total survival times were not statistically significantly different. This may have been due to the initially high activity levels of the EC rats, having the net effect of exposing them to predation for much more of the time than the IC rats. The survival advantage held by the EC rats after day 15 lasted until day 40 in these studies. After 40 days in a common environment, it is likely that the brain differences between the groups had diminished substantially. The subsequent lack of differential survival in animals previously housed in EC and IC is, therefore, difficult to interpret (see the section on persistence of EC-IC effects in the following chapter).

In the second study (Renner, 1987b), EC rats took significantly less time than did IC rats to escape from an arena under simulated predation on the first day in that arena. For EC and IC rats given opportunity to explore the arena for 10 minutes on each of two days prior to being subject to the simulated predation on day 3, the differences were larger than for rats tested on day 1. These behavioral EC-IC differences are not only clearly functionally important, but their increased magnitude resulting from two brief opportunities for exploration supports the conclusion that EC and IC rats acquire different information during exploration as a result of their different behaviors.

Just as knowledge of the particulars of functional neuroanatomy leads to specific questions about changes in behavior concomitant with alterations in that anatomy, it is possible that knowledge of changes in particular behaviors or in behavioral organization could lead to asking specific or even different questions than we ask at present about underlying changes in the structure or function of the nervous system concomitant with behavioral change.

Chapter 4:
Boundary Conditions for Environmental Effects

As soon as the first reports concerning neural plasticity related to experience became known, an understandable concern was expressed over whether these findings would obtain in different circumstances than the exact situation present in the Berkeley laboratories. In the ensuing quarter of a century, multiple laboratories have investigated EC-IC effects in other strains of rats, in rats of both sexes and several ages, and in other species. The nature of enrichment and impoverishment as relative conditions has also been investigated by variation of environments along a continuum from extreme impoverishment to attempts to "superenrich" the environment. The time required for the appearance of significant neurobiological changes has been a topic for investigation as well. Each of these areas of inquiry will be reviewed in this chapter.

Generalizability Among Strains of Rats

Krech, Rosenzweig, and Bennett (1960) addressed the question of generalizability of EC-IC effects in the first report of neural responsiveness of enriched and impoverished environments, by comparing responses to differential environments of several different strains of laboratory rats, finding only minor variation across strains. The S_1 and S_3 strains employed in the original studies in the Berkeley laboratory are descended from the maze-bright and maze-dull strains of Tryon (1940); both strains show clear EC-IC effects. Similarly, Ferchmin, Eterovic, and Levin (1980) found typical EC-IC effects on cerebral weight in rats specifically bred for high and low active avoidance ability. In work with homozygous Brattleboro rats, Greer, Diamond, and Murphy (1982) have also demonstrated environmentally induced plasticity. Rosenzweig and Bennett (1977) report significant effects of enrichment on both weights of brain regions and cholinergic measures in Fischer rats.

Multiple investigations, some predating the discovery of neural plasticity, have also shown experiential effects on problem-solving to be general across strains, including Sprague-Dawley (Woods, et al., 1961), Wistar (Bingham and Griffiths, 1952), Long-Evans (Brown, 1968), and Holtzman (Ough, et al., 1972).

Cross-Species Generalizability

Studies that use laboratory rats exclusively leave open the question of whether the results may be restricted to this species. In a more general sense, different species evolved to fill particular niches in the natural world, and each species is therefore specialized; laboratory rats have certainly been subject to laboratory selection, both practical (e.g., resistance to disease) and inadvertent (e.g., failure to breed colony animals that are exceptionally difficult to handle). By studying several species' reactions to an experimental manipulation, it becomes possible to separate effects that are general in character from those that are due to unique qualities of a single species.

Studies of environmental enrichment and impoverishment have been carried out with several species to date, and differences in brain chemistry and anatomy similar to those described above from the voluminous literature derived from work with rats have been found in several families of rodents; we know of no unsuccessful attempts to discover experientially-related plasticity in systematic research with any species. Several studies on non-rodent subjects suggest fruitful ground for further investigation, as described below.

Mice

Most major aspects of neurobiological responses to differential environments found in rats, both anatomical and chemical, have also been described in *Mus musculus*; mice are being used more frequently in studies that involve enriched and impoverished environments, both as a primary independent variable and as an experimental manipulation in the service of studying other types of problems.

La Torre (1968) replicated the original Berkeley experimental paradigm using two strains of mice (strains C57BL/Crgl and A/Crgl of *Mus musculus*, from the same family as rats, *Muridae*), and found the results to be comparable for all measures taken. Brain weights were significantly higher in ECT than IC for both strains, as were total acetylcholinesterase and cholinesterase. Henderson (1970, 1973) replicated this finding (albeit in a complex environment without frequent changes of stimulus objects), using whole-brain weight in 6 inbred strains of mice. Collins (1970) failed to find EC-IC differences in two strains of mice bred for high and low brain weight; the reasons for the discrepancy between those results and those of other investigators remains obscure.

Cummins, Livesey, & Bell (1982; 1983) found increases in forebrain weight in EC averaging 5% over several experimental durations in mice. They also reported significant increases in cortical thickness in mice kept in an enriched condition from weaning to age 30, 40, or 50 days but not to 70 or 100 days (although small sample sizes for the older mice impair the interpretability of these data). Warren, Zerweck, and Anthony (1982) reported that mice, assigned to EC or IC at 600 days of age for 150 days, showed an increased percentage of cortical cells containing high levels of RNA. Cordoba, et al. (1984) found EC-IC differences in protein content of the cerebral cortex comparable to those found in rats.

One apparent difference in the responses of mice and rats to differential environments was the lack of slow-wave sleep increases in EC mice (Gutwein and Fishbein, 1980a, 1980b), whereas such differences from IC have been found in EC rats (Tagney, 1973). This is one of the few replicated findings indicating that significant apparent species differences in EC-IC effects, and is difficult to interpret in the absence of other meaningful discrepancies.

Other types of problems that are increasingly being studied with mice include the applications of environmental enrichment to problems of aging (e.g., Kubanis, Zornetzer, and Freund, 1982) and recovery from brain trauma (e.g., Goodlett, Engellenner, Burright, and Donovick, 1982). (Applications of results from studies of enriched and impoverished environments are discussed in Chapter 6.)

Gerbils

Rosenzweig and Bennett (1969) studied the mongolian gerbil (*Meriones unguiculatus*, from the family *Cricetidae*) and its neural plasticity. In most respects results for gerbils resemble quite closely those obtained for rats, even more closely than the results of studies conducted with mice resemble those with rats. The EC gerbils exceeded IC gerbils in cortical weight and in ratio of cortical to subcortical weight. In addition, the ratio of adrenal gland weight to body weight did not differentiate EC from IC gerbils;

likewise, the adrenal weight/body weight ratio is not consistently reported to be different between EC and IC rats. EC-IC differences in AChE were equivalent to those found in rats, but these experiments revealed significant differences in ChE concentration only occipital cortex for gerbils. Comparison of the ratio of concentrations of the two enzymes (ChE/AChE), which shows significant EC-IC differences in occipital cortex and total cortex, revealed differences in the gerbil only in occipital cortex. As Rosenzweig, Bennett, and Diamond (1972b) have reported that in long-duration experiments the EC-IC difference in ChE is largely due to a decrease in this enzyme in IC (when compared to SC), it is possible that this rat-gerbil difference in environmental responsiveness indicates less response by gerbils to impoverished environments; this possibility would require further investigations.

Cheal, Foley, and Kastenbaum (1984, 1986) report significant effects of very brief (one hour per month) experience in an enriched environment on both on the behavior (1984) and biology (1986) of gerbils. It should be noted, however, that sex differences in response to this brief enrichment (producing opposite effects for some variables, such as skeletal growth) bespeak caution in interpretation of these results.

Ground Squirrels

Two species of ground squirrels, (both the Belding's ground squirrel [*Spermophilus beldingi*] and the golden-mantled ground squirrel [*Spermophilus lateralis*], of the suborder *Sciuromorpha*) have been used in studies of neural and behavioral response to environmental manipulation (Rosenzweig, Bennett, & Sherman, 1980; Rosenzweig, Bennett, Alberti, Morimoto, & Renner, 1982; Rosenzweig, Bennett, Renner, & Alberti, 1987). Both chemical and anatomical measures in sciurid brains are more variable than in rats (probably due to the selection pressures to which rats are exposed in the laboratory, discussed above, and that have not affected the wild squirrels). Nevertheless, significant differences in cortical weight and nucleic acids as a function of environmental treatments, have been found in these species of ground squirrel. Furthermore, these differences show patterns of regional specificity, in both presence or absence of differences and magnitude of differences, that are in most instances comparable with those found in rats. The only evidence counter to this general finding (Renner and Rosenzweig, 1987) employed individual squirrels in an enriched condition, and both these subjects and those in IC displayed clear evidence of isolation stress, a syndrome not seen in laboratory rats in isolation of durations typically used in EC-IC experiments.

In these studies, squirrels housed in the laboratory in enriched conditions have also been found, for most measures, to have cortical weight and nucleic acid content not significantly different from same-age squirrels caught in locations nearby to where the mothers of the laboratory subjects were trapped (Rosenzweig, et al., 1980, 1982, 1987).

Cats

As discussed in Chapter 3, Cornwell and Overman (1981) and Wilson, Warren, and Abbott (1965) found that environmental enrichment yielded advantages on Hebb-Williams maze problems in cats, although Wilson, et al. did not find EC-IC differences on alternation or active avoidance learning. Wilson, et al. (1965) also reported decreases in timidity towards humans and deficiencies in active avoidance learning in handled cats when compared to isolated controls. Concerning neurobiological effects of enriched and impoverished environments, Beaulieu and Colonnier have recently reported several studies with domesticated cats in which they replicated the EC-IC differences found for rats in neuronal size and density (1985), synaptic density (Colonnier and Beaulieu, 1985), and numerical increases in some types of axonal boutons (Beaulieu and Colonnier, 1986). These results, although not comprehensive, indicate substantial verification of rodent EC-IC studies in a species of carnivore, *Felis domestica*.

Primates

Monkeys reared in a colony conditions very similar to laboratory environmental enrichment in rats (including social housing and numerous large and small stimulus objects) were superior to isolation-reared monkeys on complex oddity tasks, but not on simple discrimination or delayed-response problems (Gluck, Harlow, and Schiltz, 1973). In addition, as described in Chapter 2, Floeter and Greenough have demonstrated plasticity of Purkinje cells of the cerebellum in Japanese macaques, *Macaca fascicularis*, as a function of differential environments (1978, 1979). Measures of the cerebral cortex were not reported for these subjects. These results do indicate that there may be neural plasticity in response to environmental manipulation in primates as well as rodents.

Although it is rare in the 1980s that the cross-species generality of neurobiological plasticity is seriously questioned, further comparative research could open new avenues for potential applications of these phenomena, through better understanding of the specific character of different species' reactions to differential experience and the details of methodology for producing EC-IC differences in different species. By studying these patterns of cross-species commonalities and differences, it may be possible to gain further insights into mechanisms of neural plasticity, learning, and memory.

Sex Differences in Environmental Effects

Although the majority of studies of neural plasticity in response to environmental manipulation have been carried out with male rats, the earliest reports (Krech, et al., 1960) addressed the issue of cross-sex generality of EC-IC effects, by including groups of females run in parallel to studies done with males. The female rats showed EC-IC differences in chemical measures of the cholinergic system equivalent in magnitude to those found for males.

Further studies have investigated the potential role of male-female differences in EC-IC effects by examining responses to environmental manipulation in female rats. In studies on pregnant and nonpregnant female rats, two studies reported significant differences in depth of occipital cortex between nonpregnant EC and IC females (Diamond, Johnson, and Ingham, 1971; Hamilton, Diamond, Johnson, and Ingham, 1977). The reports differ, however, on environmental effects in the pregnant rats. While the earlier study (Diamond, et al., 1971) reported that EC-IC differences in cortical depth in pregnant females did not achieve significance, in the later study (Hamilton, et al., 1977) the differences were significant. Both studies attributed the diminution of EC-IC differences in pregnant rats to an increase in cortical depth of the pregnant ICs, relative to EC. A third study (Pappas, Diamond, and Johnson, 1978) reported results of combined ovariectomy and differential environments: sham-operates developed EC-IC differences in cortical thickness in the occipital region, ovariectomized ECs showed increases relative to sham-operate ECs in motor (but not occipital) region thickness, and ovariectomized ICs had significantly thicker cortices than sham-operate ICs in somesthetic, motor, and occipital areas. Although Pappas, et al. discuss these findings as though ovariectomy and associated alterations of endocrine function in female rats prevented the occurrence of EC-IC effects on cortical thickness, no direct comparisons of brain measures between ovariectomized EC and IC were reported; in fact, ovariectomized EC had higher mean thickness than ovariectomized IC in all four occipital dimensions reported. Although these data suggest that there may be a sex difference in responsiveness to environmental stimulation mediated by endocrine factors, none of the results for females were compared statistically to brain effects in male rats. The lack of direct male-female comparison in these studies leaves this question unresolved.

Direct comparisons of EC-IC effects between males and females have been few. Analysis of the possible existence of sex differences in effects of differential environments is complex; results are mixed, with some studies reporting differences between the sexes and others an absence of such differences. Interpretation of these findings is further complicated by the lack, in most studies, of direct statistical comparison between behavioral measures of males and females. For example, Rosenzweig and Bennett (1977) conducted parallel experiments with males and females rats of a hybrid (S_1 X Fischer) strain in enriched and impoverished conditions, with somewhat mixed findings. While the ratio of cortical to subcortical weight showed an EC-IC difference of 5.6% in males ($p<.01$), it was only 3.3% in females ($p<.01$). A similar apparent difference was found for the cholinesterase-acetylcholinesterase ratio when cortical samples were compared to the rest of the brain: the EC-IC difference was 7.6% in males ($p<.001$), 3.5% in females ($p<.01$), although the ChE/AChE ratios were much closer in cortical measures (11.0% for males, 11.3% for females; both $p<.001$). No statistical tests of these sex differences in size of effects were performed.

Juraska (1984a, 1984b) performed direct comparisons between males and females of EC-IC effects on dendritic branching in several layers of the occipital cortex. She reports that females show smaller environmental effects on dendritic length and branching in layer III pyramidal cells, for both apical and oblique dendrites. A similar but weaker trend was evident in layer IV stellate neurons, but layer V pyramidal neuronal response to environmental manipulation showed no sex differences. In the dentate gyrus of the hippocampus, males showed little evidence of EC-IC differences in dendritic branching, while EC females have significantly greater branch length than IC females, producing a statistically significant sex-by-environment interaction in dendritic response (Juraska, Fitch, Henderson, and Rivers, 1985). In the corpus callosum, although Juraska and Meyer (1985) reported enrichment-induced increases and did not observe sex differences in gross size, more detailed investigation (Kopcik, Juraska, and Washburne, 1986) revealed differences for females, but not for males, in number of myelinated axons; both sexes had higher numbers of unmyelinated axons in enriched environment groups.

There have been findings of sex difference in response to brief enrichment from gerbils as well (Cheal, Foley, and Kastenbaum, 1984, 1986): Both males and females in these studies showed enrichment-related increases in the size of the ventral (scent-marking) gland, but males with brief enriched experience showed increases in skeletal growth whereas enriched-experience females showed decreases. Furthermore, enriched males showed increased susceptibility to seizures, while there was no effect of enrichment on females' seizure rate. Although the methodology in these studies is not precisely parallel to that of other types of enrichment studies (the timing and particulars of the enrichment, as well as the repeated measurements taken from a single group of subjects), the sex differences in apparent response to environmental manipulation may provide clues as to the endocrine mediation of some EC-IC effects.

The available behavioral evidence presents a contradictory picture about sex differences in effects of experience. While Woods, Ruckelshaus, and Bowling (1960) and Caul, Freeman, and Buchanan (1974) report sex-by-environment interactions on behavioral tests, the types of interaction are not concordant: Woods, et al. report a larger EC-IC difference in Hebb-Williams performance for females, Caul, et al. report that male ICs show a greater degree of Pavlovian response suppression than females, while male and female ECs are equivalent, indicating larger EC-IC differences in males. Joseph (1979) reported that males were superior to females in a symmetrical maze similar to the Hebb-Williams mazes, in parallel to but not interacting with an EC-IC difference. Juraska, Henderson, and Muller (1984), however, report that they found EC-IC difference in radial maze performance, without sex differences or sex-by-environment interactions. With the exception of the Juraska, et al. (1984) study, which reported an internal replication, each of these reports consists of results from a single experiment. There is, therefore, no obvious basis for resolving these widely discrepant results without further empirical work.

In a separate line of investigation, Sackett (1972), in a retrospective examination of the several studies of mother-infant separation and subsequent isolation in rhesus macaques (*Macaca mulatta*), reported that gross behavioral abnormalities were less common in female than in male isolates. While it is not at all clear that there are common mechanisms involved in early social isolation and differential environments, sex differences from these investigations may provide material for hypotheses concerning sex differences in the behavioral effects of differential environments.

Although it seems clear that there may be sexual dimorphism in some aspects of neural responses to enrichment and impoverishment, the exact character and potential functional significance of this dimorphism remains unclear.

Superenrichment and Extreme Impoverishment

As discussed in Chapter 1, semantic issues can become problematic in attempting to describe experimental manipulations of the environment. The terms "enriched" and "impoverished" are not definitive, but rather label the environments relative to a typical laboratory housing condition, which is usually social housing in empty cages. Rather than the only two options in a dichotomy, the enriched and impoverished conditions used in many experiments are at points along a continuum of environmental complexity comprised of several variables, intended to be different enough with regard to those variables that biological and behavioral altered by them will show measurable differences.

Holson (1986) and Rosenzweig, et al. (Rosenzweig, Bennett, and Diamond, 1972c; Rosenzweig and Bennett, 1978) have employed highly similar environmental conditions (i.e., sand or soil substrate, outdoor temperature and lighting, large enclosure, and stimulus objects available in the environment), and described these environments, respectively, as enriched (EC) and Semi-Natural Environments (SNE). Each of these environments is, in fact, enriched relative to typical laboratory housing; it is only because the degree of environmental complexity can affect brain measures that this semantic issue becomes problematic.

There have been occasional attempts to create laboratory environments more enriched that the typical EC conditions, with somewhat mixed results. Ferchmin, Eterovic, and Caputto (1970) reported studies employing an environment similar to the Berkeley EC, but using several different sizes and shapes of cages, and a mixed-age group of rats. In direct comparison between EC and this "Ferchmin EC" (FEC), Bennett, et al. (1974) found that FEC rats had cortical weight and AChE differences from IC somewhat greater than those in EC rats, but ChE concentration and ChE/AChE ratios somewhat less than those of Berkeley EC rats. Kuenzle and Knusel (1974) reported comparisons between brain measures of animals in enriched conditions identical to those used in the Berkeley laboratories and a "superenriched" condition (SEC), a set of linked cages stocked with a rotating population of 70 subjects at a time as well as innumerable stimulus objects, mazes, and additional complexities varied daily. Davenport (1976) reported some evidence of better success in applied use of differential environments with superenrichment than with standard EC treatment (discussed in Chapter 6). Using similar procedures, however, Renner, Rosenzweig, Bennett, and Alberti (1981) found the differences between SEC and EC brain measures too small to reliably attain statistical significance. Renner, et al. did show that the relationship between measures of cortical weight and repeated errors in the Hebb-Williams mazes to be reliable, both within and between environments; this lends indirect support to the efficacy of superenriched environments.

Seminatural conditions (SNE), attempts to simulate wild conditions in controlled experimentation, have been employed as well. In four experiments, Rosenzweig and Bennett (1978) found that brain values of rats from SNE significantly exceeded those of EC rats for several measures, including weight of occipital cortex (5.0%, $p<.001$), total cortex (3.1%, $p<.001$), and cortical/subcortical weight ratio (3.1%, $p<.001$). Similar results were obtained for acetylcholine activity in these areas. These findings, although they indicate that laboratory enrichment effects may not typically be limited by ceiling effects, are not entirely consistent with the results from studies of ground squirrels (discussed above), where wild-caught same-age *S. lateralis* and *S. beldingi* did not reliably exceed laboratory-born squirrels with enriched experience (Rosenzweig, et al., 1980, 1982, 1987).

There is evidence that, just as laboratory enriched conditions are not maximally enriched, laboratory impoverished conditions are not maximally impoverished. Krech, Rosenzweig, and Bennett (1966) conducted studies with isolates or paired animals in extreme impoverishment (housing in sound-attenuating chambers under diffuse light, with measures taken that the subject not see the experimenter during routine maintenance). Their results indicated little support for the protective effect of paired housing in these conditions; the extremely isolated groups (paired and singly housed) were not consistently different from one another, and both groups had significantly lower weights of brain regions than rats housed in the usual SC situation, which were lower than the EC group. Although the more-common IC groups were not included, making direct comparisons of effect size impossible (a complication shared by most studies of possible sex differences in EC-IC effects, discussed above), the EC groups exceeded the extremely impoverished groups by somewhat more than typical EC-IC differences for most measures.

Persistence of Effects

It would seem plausible that, if plasticity is a general principle of neural function, the changes in neural anatomy and chemistry induced by enriched or impoverished experience would be reversible given proper environmental conditions. Available evidence supports this hypothesis. Rosenzweig, Krech, Bennett, and Zolman (1962) examined the reversibility of EC-IC effects by moving one group from IC to EC midway through an 81-day experimental period. Weight of a posterior section of the neocortex (including occipital and sensory samples), total cortex, and cortical/subcortical weight ratios for this group were intermediate between EC and IC; although cholinesterase values were not significantly different among the groups, the means generally followed this pattern; the few exceptions were in cases where values from the switched group was equivalent to those from EC. Bennett, Rosenzweig, Diamond, Morimoto, and Hebert (1974) systematically investigated persistence of environmentally-induced effects following moving the EC subjects to IC, finding that measures of occipital cortex weight and AChE concentration in the EC-to-IC switched group were still significantly different from the IC group 21 days following the switch. By 32 days post-switch, however, differences between the groups were no longer significant. Bennett, et al. also found that brain changes induced by 80 days' EC-IC differential housing are more persistent that those induced by 30 days' differential housing.

These results, that environmentally induced neural differences can be attenuated by subsequent environments, are corroborated by other work on the anatomy and physiology of neurons. Ferrer (1983) examined dendritic spines in enriched and impoverished animals after a six-month period in social housing, finding that EC-IC differences induced in number of spines by differential housing did not persist through this delay. Green and Greenough (1986) report that the differences in synaptic transmission reported in Chapter 2 are not found in EC subjects rehoused in IC for 3-4 weeks prior to removal of hippocampal slices.

Time Constraints on the Appearance of Neurobiological Changes

Although Ferchmin, Eterovic and Caputto (1970) reported small, transient increases in RNA/mg after 4 days in EC, in experiments of longer than 16 days in duration the RNA/mg differences between EC and IC groups were very nearly zero. Total RNA in the cortex also showed increases in as little as 4 days of enriched environment (Ferchmin, et al., 1970), and increases in total cortical RNA have been found in all experiments of 4 days or longer duration (Bennett, 1976). More recently, Ferchmin and Eterovic (1986) have provided evidence that four daily 10-minute sessions in an enriched environment are sufficient to produce significant increases in RNA content of the occipital cortex.

Rosenzweig and Bennett (1978) examined the relative rapidity with which the environmentally induced anatomical changes can take place, and found significant EC-IC differences in the weight of occipital cortex and total cortex first becoming significant after four days' exposure to differential environments. These findings are supported by other work: Using mongolian gerbils (*Meriones unguiculatus*) as subjects, Cheal, Foley, and Kastenbaum (1984, 1986) provide evidence that one hour monthly of enrichment in an outdoor environment leads to significant differences from controls in several biological variables (including cranial volume).

Many of the neurobiological effects of differential environments are in the same direction and are of similar magnitudes, in the range of five percent. This seems to have contributed to the practice of different investigators using different measures of neurobiology (e.g., weights of cortical regions and neurochemical measures) almost interchangeably, as though they were in fact measuring a single event. In trying to reach an understanding of the nature of the effects of differential experience, which in turn may suggest avenues for application and potential mechanisms of these differences, it is important to point out that the anatomical and biochemical changes are not isomorphic. More detailed examination of the individual measures, however, makes it plain that these measures are indications of many types of change occurring in the brain. Rosenzweig, Bennett, and Diamond (1972b) report that differences in cholinergic enzymes increase in magnitude as duration of differential housing increases, while there is some evidence that cerebral weight differences decrease (Bennett, Rosenzweig, and Diamond, 1970). Walsh, et al (1973) found that differences in cerebral dimensions develop more slowly than similar-magnitude changes in brain weight or biochemistry. Similarly, Sirevaag and Greenough (1986) report a statistical analysis in which different measures of neural microstructure discriminate EC from SC and IC than discriminate IC from SC and EC. Collectively, this evidence makes a clear case that oversimplification of the way in which neural effects of differential environments are described creates a danger of misclassifying a complex range of phenomena as a single event.

Chapter 5:
Causes of EC-IC Brain Differences

Enriched and impoverished environments are demonstrably different in several respects. Likewise, subjects with experience in differential environments are different in several types of behavioral measures. Of the many environmental characteristics manipulated as independent variables, and the many subject characteristics shown to be altered by enriched and impoverished laboratory environments, there are few that have not been singled out at one time or another and proposed to be *the* critical difference between EC and IC animals, and therefore the cause of the brain changes that result from differential experience. This section will examine several of the possibilities that have been suggested, examining the data relevant to each.

A Comment About Levels of Analysis

Although two of the hypotheses outlined below, those concerned with the respective roles of play and object interactions, include elements contained in hypotheses advanced before, their value is apparent when it becomes clear that each is at a different level of analysis than the explanations advanced previously. Maturation, stress, endocrine or neurochemical effects, social stimulation, and learning and formation of memory are proposed as intervening variables, labelling some hypothetical internal event or state which is then held responsible for the observed brain and behavioral changes. Play and object interaction, on the other hand, are concrete types of behavior that are suggested to be necessary for the development of the typical EC-IC brain and behavioral differences. This latter approach is advantageous in that is lends itself more directly to an empirical approach. Play and object interaction can be directly measured (once operationally defined in specific behavioral terms), whereas stress and maturation, for example, are much broader constructs, and can be measured only indirectly. Thus, while both the intervening variables approach (mechanism-oriented) and the more specific behavior types approach (activity-oriented) share the liability that neither level of analysis lends itself to conclusive support, the activity-oriented hypotheses do lead to specific, falsifiable predictions about what would be observed in the behaviors of subjects in these conditions.

Handling and Locomotion

The earliest reports of neural plasticity induced by differential environments included control experiments to examine possible roles played by handling and differential locomotion in these effects (Krech, Rosenzweig, and Bennett, 1960), and additional tests were reported by Bennett, et al. (1964) and Zolman and Morimoto (1965). No support was found for the hypothesis that differential locomotion and differential handling played roles in the differences found in the cholinergic system.

Maturation

It has been suggested that brain differences between EC and IC subjects are caused by different rates of maturation (Cummins, Livesey, Evans, & Walsh, 1977). The simplest form of this hypothesis proposes that the increases in brain weight shown by enriched condition subjects are the result of accelerated maturation, i.e., EC rats were doing their growing sooner, but that the IC rats would do the same growing later. This implies that, if left in the differential environments long enough, the differences between the groups would disappear. Although it is true that for some measures the EC-IC differences go through a peak and then decline (see Bennett, Rosenzweig, and Diamond, 1970), the differences do not disappear. Differences in cortical weight between EC and IC groups did not disappear in 500 days of differential housing in experiments carried out by Cummins, Walsh, Budtz-Olsen, Konstantinos & Horsfall (1973).

This hypothesis has, at its core, an assertion that environmental enrichment has its effect by altering the temporal pattern of the normal course of brain development. This assertion was undermined by the results of a study by Riege (1971). In this experiment, rats who were differentially housed for the first time at approximately 10 months of age (middle-aged for rats) showed significant differences in brain measures. Although the brain differences observed in this study were somewhat less than those typically found in experiments with weanlings, they were within the normal range of study-to-study variability. Behaviorally, Doty (1972) found Hebb-Williams performance differences in rats put into differential housing at 300 days of age, about the same age as the subjects in Riege (1971). More detailed study, using rats not given differential experience until 600 days of age, showed large differences in higher-order dendrites (Connor, Melone, Yuen, and Diamond, 1981). Green, Greenough, and Schlumpf (1983) found similar results for rats placed in EC or IC at 450 days of age and kept there for 45 days. Likewise, Warren, et al. (1982) report significant experiential alterations of both behavior and brain weight in adult mice first given differential experience from 600 to 750 days of age.

The evidence that the adult rodent brain remains plastic notwithstanding, there are limitations of plasticity in adult and old animals that are not present in the young. Fiala, Joyce, and Greenough (1978) demonstrated that dendritic field size and complexity of dendritic branching in hippocampal dentate gyrus were increased compared to impoverished littermates in rats housed in enriched conditions at weaning, but were not affected in rats housed socially until 145 days of age prior to being placed into differential environments. Sharp, Barnes, and McNaughton (1985) report that the long-term enhancement in cells of the dentate gyrus from enriched experience rats (Sharp, McNaughton, and Barnes, 1983, 1984) decays more quickly in extremely old rats (32 months of age) than in adults (14 months).

Uphouse (1980) has argued that the apparent decrease in plasticity in older rats constitutes support for the hypothesis that EC-IC differences are due in part to differences in maturation rates for the groups, but the logic of this argument is unclear. The critical issue in the question of different rates of maturation is the presence or absence of significant plasticity in adult and aging animals, not its magnitude relative to that in young animals. While experientially induced brain changes may be more limited in adults than in weanlings, the fact remains that the adult brain shows considerable plasticity. This casts serious doubt upon the adequacy of the differential maturation hypothesis for explaining the brain differences observed between EC and IC animals. Just as bones calcify with aging, normal developmental processes may lead to decreased responsiveness to environmental influences, by lessened response in each responding cell, responses by fewer cells, or by some combination of both mechanisms. Even if the brain does experience reduced plasticity, this does not alter the empirical fact that the brain of the rat remains plastic in adulthood and old age. One could not easily argue that for adult or old animals, the cerebral changes induced by environmental enrichment represent enhanced maturation.

Stress

Isolation is widely seen as a stressor, producing evidence of deleterious effects such as caudal dermatitis, aggressiveness, and enlargement of the adrenals. On the other hand, the enriched condition, in addition to being more complex and challenging than the impoverished condition, also provides more uncertainty for the animals who reside there, especially in those laboratories employing frequent changes made in the environment as part of the enrichment procedure. Stress could play a role in EC-IC effects via either scenario, but the majority of inquiry concerning the possible role of stress has focussed upon the possibly stressful effects of the impoverished environment.

The hypothesis that stress should be considered as an important variable in the EC-IC brain differences was considered by Bennett, Diamond, Krech, and Rosenzweig (1964). In addition to finding a lack of behavioral evidence of stress in IC animals, they did not observe EC and IC to differ in adrenal weight. Other empirical reports fail to provide clear support for this hypothesis; reports of environmental effects on adrenal tissues are mixed. Krech, Rosenzweig, and Bennett (1966) reported significant differences between rats housed singly in extremely impoverished surroundings and EC rats: isolates exceeded EC by 11.8% in adrenal weight. They further reported that the ratio of adrenal weight to body weight (for which IC exceeded EC by 8.9%) did not differentiate these groups. Riege and Morimoto (1970) reported similar results: The adrenal glands of ICs were significantly heavier than those of EC, but the IC group also exceeded EC group in body weight; when this was taken into account, it reduced (but did not eliminate) the magnitude of this difference. Geller, Yuwiler, and Zolman (1965) report increases in adrenal/body weight ratios for IC rats compared to either SC or EC (although their procedure of weaning pups at 19 days of age may have made their subjects more susceptible to stress). Uphouse and Brown (1981) also reported that IC have an elevated adrenal/body weight ratio, but Rosenzweig, Bennett, and Diamond (1972b) presented further evidence that the Berkeley laboratories found no differences between EC and IC rats on this measure. In adult rats, 90 days of age at introduction to differential experience, Wallace, Black, Hwang, and Greenough (1986) report that adrenal weight in EC rats significantly exceeded IC littermates after 10 days in differential environments, and that differences of equivalent magnitudes were also found in animals housed differentially for 30 or 60 days. We know of no obvious explanation for these discrepant results, although Devenport, Dallas, Carpenter, and Renner (1986) report that adrenalectomy-induced cortical growth is independent of EC-IC brain changes. This may imply that adrenal influences on brain development can co-occur with, but not play a role in, brain changes related to differential environments.

The magnitude of EC-IC cerebral effects was not changed when both EC and IC rats were subjected to chronic stressors (Riege and Morimoto, 1970). Further, when stress alone was the independent variable, brain changes of the type found in typical EC-IC experiments were not produced. In experiments where stress is not an independent variable, neither EC nor IC rats show external evidence of stress (e.g., caudal dermatitis), and behavioral observations do not reveal other signs of stress such as difficulty in handling.

Endocrine System Alteration

Rosenzweig, Bennett, and Diamond (1972a) examined the possibility that the different conditions to which the animals were exposed somehow altered the hormonal balance of animals in one or both conditions in such a manner as to cause the observed effects of differential experience. Neonatally hypophysectomized rats were placed at weaning into standard EC and IC conditions. Although the biological development of these rats was far from normal, EC-IC differences of approximately normal percentage magnitude were found. As discussed above, Wallace, et al. (1986) have recently provided evidence that EC-IC effects on adrenals, if they exist at all, may be opposite in young rats and adults.

While it can safely be concluded from these data that normal levels of pituitary hormones or hormones regulated by tropic pituitary hormones are not essential in order that the biological events leading to the observed brain effects of environmental enrichment take place, it is not reasonable to infer that endocrine events play no part in enrichment effects in normal animals. The role of the endocrine system and arousal in modulating other psychobiological processes is well documented (see, for example Rosenzweig and Bennett, 1984, for discussion of the role of modulatory processes in memory formation). Therefore the possibility of modulatory roles of the endocrine system in the effects of environmental enrichment is a subject deserving further investigation.

Neurochemical Alteration

Norepinephrine (NE) has been implicated in several phenomena possibly connected with EC-IC effects: arousal, learning and memory (Kety, 1970); neural plasticity (Kasamatsu, Pettigrew, and Ary, 1981); and investigatory behavior (Flicker and Geyer, 1982). Recently, two laboratories have reported that injection of 6-hydroxydopamine (which depletes brain NE) reduces EC-IC effects, suggesting that NE is causally implicated in these effects (Mirmiran, Brenner, and Uylings, 1983; O'Shea, Saari, Pappas, Ings, and Stange, 1983; Pappas, et al., 1984) Another study, however, provides evidence that NE depletion decreased both investigatory behavior and home cage activity, in addition to alteration of

EC-IC brain effects (Benloucif, Rosenzweig, and Bennett, 1984). This evidence weakens the argument that EC-IC effects depend directly on NE, and further suggests that decreases in EC-IC effects are a secondary consequence of altering the animals' behavior patterns. Coyle and Singer (1975a, 1975b) reported similar findings in a study of EC-IC effects in rats prenatally exposed to imipramine; compared to saline-treated controls, imipramine-treated rats showed smaller EC-IC differences, but also exhibited lower levels of interaction with the environment. This suggests again that decreases in EC-IC effects are secondary to behavioral changes in the drug-treated subjects.

Pearlman (1983) has provided evidence that injection of clonidine or methyldopa immediately after daily two-hour experience in EC rats reduces the increase in cortical weight that usually accompanies enriched experience. Pearlman suggests that normal functioning of the adrenergic system, due to its support of REM sleep, is essential for the formation of EC-IC brain differences. This study, however, does not establish that the behavior of the drug-treated subjects was equivalent to that of the saline-injected control EC subjects. That is, the daily injections may have reduced the interactions of the rats with the enriched environment; observations of subject behavior, including interaction with the environment, were not reported in this study. Since interaction of the subjects with the environment has been shown to be critical to the development of EC brain effects (Ferchmin, Bennett, and Rosenzweig, 1975, discussed below), a complete interpretation of these results is not possible.

Social Stimulation

Although there is a considerable literature on social isolation in rodents and other animals, with or without concomitant sensory deprivation, the isolation most commonly employed in EC-IC experiments is of shorter duration and lower degree than in isolation experiments. If experience is conceived as scalable on a continuum, EC and IC are separated far enough to generate clear, observable differences and are intended as models of naturally occurring variations of experience; severe isolation and stress-by-overload (cf. Lindroos, Riittinen, Veilahti, Tarkkonen, Multanen, and Bergstrom, 1984) are more extreme examples, toward the ends of the experiential continuum beyond, respectively, IC and EC. Bennett, et al. (1964) found no evidence of isolation stress in ICs, and Krech, et al. (1966) reported that even extreme isolation in the durations typically used in EC-IC studies produce no evidence of isolation stress. The phenomenon of behavioral pathology in isolation apparently plays little or no role in EC-IC differences. In addition, Einon and Morgan (1977) and Einon, Humphreys, Chivers, Field, & Naylor (1981) provide evidence that there critical periods in formation of isolation stress, and there is much evidence, discussed above, that there are no critical periods in the formation of EC-IC brain and behavioral differences.

The difference in social stimulation is an obvious difference between EC and IC, and several variations of the possibility that social interactions play a crucial role in EC-IC differences have been suggested. Welch, Brown, Welch, and Lin (1974) argued that social housing created significant brain differences from IC housing and was therefore responsible for EC-IC differences. While social stimulation may have some role in EC-IC effects, it has been established that it cannot account for these effects. In an investigation of the contribution of social factors to the brain effects of enrichment and impoverishment effects (Rosenzweig, Bennett, Hebert, and Morimoto, 1978), brain measures of groups of rats housed identically to EC but without toys in the cages were compared with those of rats from standard EC and IC. Group-housed rats without toys did not develop brain characteristics equivalent to EC rats; most measures were between IC and EC values. Social stimulation alone is therefore not sufficient to produce changes in brain anatomy equivalent to those brought about by enriched experience. The inanimate environment, therefore, makes important contributions to whatever aspects of the environment's complexity are instrumental in inducing the brain changes that are observed.

S.-Y. Chang (1969) produced significant brain differences in rats housed in EC for two hours per day and given pre-session injections of methamphetamine. During her experiments, she observed lower levels of social interaction in EC rats given methamphetamine than in EC rats injected with saline, and a greater EC-IC difference in methamphetamine-injected groups than in saline groups. From these observations, Chang asserted that social interaction is not necessary in producing EC-IC brain effects. The data upon which this conclusion is based, however, consist only of a negative relationship between magnitude of EC-IC effect and amount of social interaction observed. To make an inference concerning causes of EC-IC effects from this negative correlation overreaches the data.

Renner and Rosenzweig (1987) have reported upon attempts to use the supposedly nonsocial golden-mantled ground squirrel (*S. lateralis*) as a model for experimental dissociation of the effects and importance of social stimulation from other differences in enriched and impoverished environments. Reasoning that a non-social animal was not subject to possible deleterious effects of isolation, we housed young squirrels in identical cages either with toys (Individuals in EC, or I-EC) or without toys (IC). The behavioral evidence collected during the experimental period of differential housing indicated clearly that these animals were experiencing isolation stress; for example, subjects from both groups exhibited extremely difficulty in handling (this was not found to be the case in this species during other experiments in which subjects were group-housed) and stereotyped pacing behavior in the home cage. This evidence renders the lack of significant differences in brain measures uninterpretable. This example of how a well-planned experiment can be seriously confounded by unexpected phenomena, the only evidence for which may be behavioral, bespeaks caution in interpreting the results of studies that do not include some form of behavioral observation to verify that the goings-on in the different environments resemble the behaviors exhibited by subjects in other experiments. If taken at face value, the results of this experiment, that a nonsocial species does not develop EC-IC brain differences without social stimulation, would have been interpreted as evidence that social interaction may be crucial for the formation of EC-IC effects, had these behavioral data not been collected.

While it would clearly be foolish to conclude that social interactions and the stimulation and opportunities for learning they afford play no role in EC-IC effects, the evidence is clear that social factors are not sufficient as an explanation of either the behavioral or neural effects of differential experience.

The Play Hypothesis

That play is an essential feature of enriched experience is a hypothesis which has been advanced recently by Robert Fagen (1981, 1982). Concerning EC-IC cerebral effects, Fagen wrote:

> "...the specific experience responsible for these changes is participation in playful social interaction, playful object manipulation, or performance of playful movements" (1981, p. 284).

Fagen predicted that play would be more frequent in EC groups than in group-housed animals. Although this seems a relatively straightforward prediction, to include social, solitary, and object-related activities under the definition of play inflates the hypothesis to near-Freudian proportions; confirmation is virtually assured, as a genuine empirical test is impossible. Any definition of play that includes interactions with objects immediately separates group-housed from enriched housed animals, because the EC subjects have a variety of inanimate objects with which to interact, while GC subjects have only food pellets. Object play, then, is only truly available to EC subjects.

If the definition of play is modified to include only social forms of play, an approach to play taken previously by Einon and co-workers (e.g., Einon, Morgan, & Kibbler, 1978; Einon, et al., 1981), it becomes a testable hypothesis, making the prediction that social play will be more frequent in EC subjects than GC subjects. Since all group-housed animals, regardless of the inanimate environment, have equal opportunity for social interactions, this hypothesis does not suffer the flaw of proposing only confounded tests.

Experimental investigation (Renner and Rosenzweig, 1983, 1986a) did not provide support for this hypothesis. No consistent differences in social interactions emerged, either in weanling (age 30-60 days) or young adult (age 90-120 days) rats in group versus enriched conditions, even in the presence of clear EC-GC differences in the weights of several cortical regions. This would seem to indicate that differences in social play are not the critical difference in behavior between group and enriched rats.

Learning and Formation of Long-Term Memory

A recurring hypothesis as to the source of brain differences between EC and IC is that the differences observed are the result of learning in EC that does not occur in IC. This idea is, in fact, a direct descendant of the hypothesis that led to the first work with enriched and impoverished environments at Berkeley -- that individual differences in brain chemistry might be related to individual variations in learning ability. The enriched and impoverished environments were originally utilized by Krech, Rosenzweig and Bennett, (1960) to provide plentiful opportunities for informal learning. The learning hypothesis is perhaps the most intuitively satisfying of all those that have been advanced; a frequent reaction of those hearing of enriched and impoverished conditions and their associated brain effects for the first time is some variant of the theme: "The environmental enrichment makes rats smarter, and this explains the brain differences, right?"

Appealing though this hypothesis may be, it is quite difficult to obtain direct experimental evidence supporting or refuting the hypothesis that learning is responsible for EC-IC differences. At best, the hypothesis can gain support from indirect sources: First, the rejection of alternate hypotheses, and second, from the collective weight of the evidence, indicating that the most reasonable and parsimonious explanation for EC-IC differences is that they are the cumulative result of many instances of learning that have taken place in EC and not in IC.

Indirect evidence in support of this hypothesis comes from several different sources: Although the early experiments using the EC-IC paradigm housed subjects in the different conditions 24 hours per day, equivalent brain changes were soon found to occur with as little as 2 hours per day of differential experience (Rosenzweig, Love and Bennett, 1968). Rats placed in EC for 2 hours a day are active for a large portion of this period, while rats that are in EC for 24 hours a day are most active immediately after the daily change of toys. Bennett, et al. (1970), found significant changes in cortical weight after 4 days' differential housing, and Ferchmin and Eterovic (1986) have recently reported that four 10-minute sessions in EC can significantly alter cortical RNA concentration. It would appear that the changed elements of the environment provoke considerable activity, and the rats in EC gain what benefits they can (e.g., information, satisfaction of curiosity, reduction of stress over the possible threat from environmental change) fairly quickly after each daily change.

Ferchmin, Bennett and Rosenzweig (1975) determined that rats must be permitted to interact directly with their surroundings if a complex environment is to have any cerebral effects. Rats placed in EC as "observers" (in a small mesh cage inside the regular EC) show brain measures and exploratory behavior similar to those of IC rats and different from those of EC rats. Renner and Rosenzweig (1986b) discovered specific behavior patterns in a novel situation in EC rats (not found in IC littermates) that must have been the result of learning which took place in the enriched environment, and had not been possible in IC.

Taken together, these studies make almost inescapable the inference that the rats in EC are deriving significant benefits from the opportunity to explore and to learn about novel objects and environmental variability. Further, the critical factor seems to be interaction rather than exposure, as has been demonstrated in numerous studies of learning.

Further evidence relevant to the hypothesis that learning and memory are important factors in producing the cerebral effects of enrichment is provided by other studies. Rats housed singly and forced to traverse a new maze daily to travel from food to water stations will show EC-like cerebral weight differences from littermates forced only to travel across an empty box from food to water (Bennett, Rosenzweig, Morimoto and Hebert, 1979). The only substantive difference between the different conditions in this study was that the maze condition placed a daily demand for learning upon the subject, while the empty box condition did not. Variables that might potentially complicate interpretation of the results of this type of experiment, such as social interaction (there was none in either condition) and locomotion are eliminated. Greenough, Juraska, and Volkmar (1979) found increased dendritic branching compared to handled IC controls in occipital cortex following multiple daily trials of Hebb-Williams maze training, similar to EC-IC dendritic differences in this area. The results of these studies would seem to make the learning and memory hypothesis much less tenuous as a possible explanation for EC-IC differences.

Additional experiments following a similar approach have been reported by F.-L. Chang and Greenough (1978, 1982). In this study, commissurotomized rats with one eye occluded by an opaque contact lens were trained on several Hebb-Williams maze problems. During the rest of each day, both eyes were unblocked. In these rats, the hemisphere opposite the "trained" eye showed significantly greater dendritic branching than the hemisphere opposite the "untrained" eye. In rats whose contact lens occluder was moved from one eye to the other each day, no hemispheric differences were observed. Brain differences found using this preparation can almost certainly be ascribed to the experiences that took place during the periods in which one eye was occluded. Clearly, then, some aspect of the Hebb-Williams training is responsible for these differences. Similar results have been obtained for motor-sensory cortex in a paw-preference task (Greenough, Larson, and Withers, 1985): Hemispheres opposite trained paws had higher total dendritic length than those opposite untrained paws.

Using different procedures, other investigators have demonstrated synaptic differences in region CA 3 of the rat hippocampus between rats given a single training session of approximately 50 minutes in a visual discrimination task and littermates performing an active control task (Wenzel, Kammerer, Frotscher, Joschko, Joschko, and Kaufmann, 1977; Wenzel, et al., 1977). These differences include number of synapses, curvature of synapses (reminiscent of the findings of Wesa, et al., 1982), and the size of the active zone of the synapse.

Based upon the evidence available at present, it would be defensible to form a working conceptualization that EC-IC differences are due, at least in part, to differential learning and resultant memory formation in the two conditions. Our hypothesis of the role of learning in cerebral differences is elaborated in part from one described in Rosenzweig and Bennett (1978), and can be summarized as follows: The additional instances of learning (in which EC animals are involved as a result of their housing condition) could induce long-term enhancement of synaptic function in some area or areas of hippocampus and neocortex of the EC animal's brain, prompting altered concentrations of some neurotransmitters and changes in synaptic morphology, size, and number, which in turn induce dendritic growth. These changes in synaptic characteristics and increases in dendritic growth result in increases in the physical size and weight of the associated brain regions. Behavioral differences observed between animals with enriched and impoverished histories would result from the combined influence of specific knowledge obtained as part of the enriched experience, as well as from an increase in information processing capacity arising from cerebral alterations. In sum, the alterations in structure and function of the brain can reasonably be seen as the cumulative effect of a myriad of small neural changes, each of which is the physical evidence of some learning. The impoverished animals, having had fewer opportunities for learning, have less evidence of it in the brain, and therefore smaller size and weight of the brain structures involved in the storage of experience.

In principle, the hypothesis that the cerebral effects of EC represent changes necessary to store long-term memory could be tested in the following way: give animals enriched experience for a limited period daily and administer to some of them an inhibitor of protein synthesis shortly before or after the daily EC period. The prediction would be that the EC animals given the protein synthesis inhibitor (PSI) would develop significantly reduced EC-IC cerebral differences in comparison with those shown by other EC animals. This design is like that which Pearlman (1983) used to investigate the possible role of the noradrenergic system in EC-IC effects. Use of this design might permit us to demonstrate complementary effects: behavioral intervention by providing opportunities to learn in an enriched environment leads to increased protein synthesis in the brain (part of which can be seen as structural changes), and somatic intervention in the form of administration of PSI treatment that prevents formation of long-term memory (LTM) also prevents (or significantly reduces) the appearance of cerebral changes typical of EC.

Bennett, Flood, and Rosenzweig considered conducting such an experiment in the early 1970s but put the idea aside because of methodological problems; further developments may make this experiment feasible now, as a brief account of the problem will show.

By the early 1970s, many investigators had conducted experiments on effects of PSIs on the formation of LTM. The results indicated that PSIs given close to the time of training could prevent formation of LTM, but there were some limitations to these experiments. The PSIs most often used in this work (puromycin and cycloheximide) had to be used at levels that were close to lethal if they were to show amnestic effects, and having to work with sick animals complicated the interpretation of experimental results. Even at these dosages, the drugs seemed to be able to prevent formation of LTM only when training had been rather weak. Bennett then explored several little-used PSIs (Bennett, Orme, and Hebert, 1972) and found that anisomycin was very suitable for research on formation of memory (Flood, Bennett, Rosenzweig, and Orme, 1973, reported the first experiments to demonstrate the amnestic properties of anisomycin). Anisomycin was not lethal when given at a dose 25 times its amnestic level. This permitted use of repeated administrations at 2-hour intervals, thus maintaining inhibition of protein synthesis at 80% or more, the degree of inhibition found to be necessary to prevent formation of LTM. Using repeated injections, longer periods of inhibition were found to prevent formation of LTM, even for strong training (Flood, et al., 1973). In principle, then, formation of memory for training of any strength could be prevented by maintaining inhibition of cerebral protein synthesis for a sufficiently long period of time.

In practice, however, certain problems arose in applying this method to investigation of the basis of EC-IC cerebral effects. In the case of rats, peripheral administration of anisomycin had to be at such large doses as to render repeated administration prohibitively costly for group experiments. Subsequent research has, however, shown that anisomycin is clearly amnestic when given intracerebrally at much lower doses, and that injections could be made on successive days through indwelling cannulas (Mizumori, Rosenzweig, and Bennett, 1985). These recent methodological advances have made this aspect of studying the role of memory-related pulses of protein synthesis in EC-IC effects feasible. At the time, however, the Berkeley group turned to research on memory formation in mice in which small doses of anisomycin administered peripherally enter the brain and are effectively amnestic.

Yet another problem also had to be considered. As we have mentioned, the stronger the training, the longer must protein synthesis be inhibited after training in order to prevent formation of LTM. This was found with one-trial passive avoidance training, when strength of footshock or duration of brief exposure to the apparatus were varied in order to obtain variations in strength of training (Flood, et al., 1973). We had no way of knowing a priori how long protein synthesis would have to be maintained following experience in the enriched environment, but seeing how rats or mice interact with many stimulus objects during a one- or two-hour daily period in EC, we feared that many hours of inhibition would be required, and we hesitated to subject animals to this treatment on a daily basis over an experiment of several weeks duration. The recent finding of Ferchmin and Eterovic (1986) that four daily 10-minute sessions in EC suffice to increase cortical RNA in young rats now suggests a feasible experiment: brief environmental enrichment of the sort used by Ferchmin and Eterovic (1986) could be employed in an experiment with anisomycin to test the hypothesis that learning and memory storage are responsible for producing the observed cerebral EC-IC effects.

Another sort of work in which Bennett and Rosenzweig are now engaged also shows related complementary effects: Brief peck-aversion training in the chick causes increases in synthesis of protein and other compounds in the forebrain, and inhibition of protein synthesis prevents formation of LTM for this training. Gibbs and Ng (1977) have shown that LTM for one-trial peck aversion training in the chick can be prevented by PSIs given within a few minutes before or a few minutes after the training trial. Memory is normal until about 60 minutes post-training, demonstrating normal acquisition and normal short-term and intermediate-term memories but failure of formation of LTM. We have replicated and extended these findings (Patterson, Alvarado, Warner, Rosenzweig, and Bennett, 1986). One-trial peck aversion training in the chick has also been reported to lead to significantly increased synthesis of proteins and other brain substances, as measured in both *in vivo* and *in vitro* preparations (Mileusnic, et al., 1980; Schliebs, et al., 1985). Increases in colchicine-binding activity were also found. Comparable increases in the rate of incorporation of [^{3}H]fucose, probably into glycoprotein, were observed (Sukuma, Rose, and Burgoyne, 1980). It is especially important in this context that Rose and Harding (1984) showed that this increased [^{3}H]fucose incorporation was abolished if the aversion-trained chicks were rendered amnesic by immediate post-training transcranial electroshock. However, if the shock was delayed for 10 minutes after training, the chicks showed both recall for the task and increased fucose incorporation when compared with either untrained but shocked or trained and immediately shocked subjects. Rose and Harding concluded that engram formation is necessary for the observed increase in fucose incorporation. Research based on these observations is now underway in the laboratories of Bennett and Rosenzweig to test the sequential dependence of the mechanisms of formation of short-term, intermediate-term, and long-term memories. In brief, Bennett and Rosenzweig hypothesize that amnestic agents that impair the formation of STM or LTM will significantly reduce the increases in synthesis of proteins and other brain compounds that are normally caused by training in chick brain.

A number of developments in the study of brain mechanisms of memory formation thus provide added support for the original hypothesis that learning and memory formation can account for the changes in brain chemistry and anatomy found between animals exposed to EC or IC. Furthermore, they now appear to make feasible experimental tests of this hypothesis.

The Object-Interaction Hypothesis

A more specific variant of the learning hypothesis is also possible. As discussed above, the effects of social stimulation and interaction cannot explain all of the differences between enriched-experience and impoverished-experience animals (Rosenzweig, et al., 1978); social grouping alone is not sufficient to produce a large part of the typical EC-IC brain differences. Previous studies have also shown that direct interaction with the enriched environment is necessary for brain effects produced by enriched experience to occur (Ferchmin, et al., 1975). In comparing rats reared in a "free environment" with toys to those reared in an identical environment but without toys -- situations reminiscent of EC and GC -- Forgays and Forgays (1952) reported significantly fewer errors on the Hebb-Williams mazes for the rats reared with toys. Finally, rats housed individually in EC cages will develop brain differences from IC littermates, but only if measures are taken to promote activity on the part of the individual in EC, this has been accomplished by injection of small doses of methamphetamine (Rosenzweig and Bennett, 1972).

These results all point to the importance of the inanimate environment and to the interactions of the animals with that environment. Taken together, they make plausible the hypothesis that object interaction and exploration are required for the occurrence of the brain effects found in enriched environment animals. Stated differently, the learning done by the EC subjects, probably about specific features of the inanimate environment, but perhaps also of information-gathering strategies, plays a critical role in the formation of brain and behavioral effects of environmental enrichment. This should in turn have a significant influence on the pattern of behaviors with which these subjects gather information about their world.

The kinship between this model of enrichment effects and the more general learning hypothesis outlined above should be apparent. They are not mutually exclusive, nor are they entirely redundant. Both hypotheses focus attention on the activities of the animals involved; both argue that what the animals <u>do</u> is at least as important as what happens <u>to</u> them.

Results from recent studies are consistent with the object-interaction hypothesis. Rats with different experiential histories do not differ in their willingness to interact with new elements of the environment (as measured by total amount of exploration and other measures), but those with enriched experience show more diversity of behavior while interacting (Renner and Rosenzweig, 1986b; Renner, 1987a).

The behavioral activity of interactions with objects in the enriched condition has been shown to lead to a relatively enduring change in behavior, that of alteration in exploratory behavior (Renner and Rosenzweig, 1986b; Renner, 1987a). "A relatively enduring change in behavior" is the usual operational definition of learning. While no definite causal link has been established between learning and what are generally considered to be positive brain changes in the enriched environment, these new findings lend further support to the hypothesis that subjects' behaviors in the different conditions are a crucial determinant of neurological consequences of environmental manipulation.

On Using Simple Explanations in a Complex World

Although much of the research directed toward elucidating causes of EC-IC effects has focussed on single variables, it would be overly simplistic to assert that one or the other variable is *the* cause of these differences. Although it is also possible that contributing factors (e.g., social interactions, behavioral response to stress, interactions with the inanimate world) operate via a common internal process such a learning, it is probable that the neural and behavioral responses to environmental manipulation are the cumulative result of compound synergistic influences of several different types of variables. What does seem clear is that varied types of support outlined here for the learning and memory hypothesis, along with the rejection of alternate hypotheses, supports the continued use of the EC-IC paradigm for providing information for the study of neural events related to learning and the storage of memory.

Chapter 6:
Applications: Enrichment as Treatment

The knowledge that manipulation of the external stimulus world can have a measurable impact on the structure and neurochemistry of the brain and upon functionally important aspects of the behavior of mammals has opened the door to a broad variety of therapeutic and preventive applications. Examples of some of the possible applications that have been attempted are reviewed in this chapter.

Enrichment for Zoos and Domestic Livestock

A major problem in zoological gardens is the relatively frequent occurrence of abnormal behaviors among captive animals. This behavioral pathology interferes with animal care, the educational mission of zoos, and attempts at breeding of captive specimens. Both these types of practical considerations and the fundamental responsibility assumed by those who capture and keep animals to treat them humanely make it essential to provide an environment that fosters normal behavior in zoo animals. Recent work has shown that creating a more complex and challenging cage environment, with natural objects, constructed artifacts or enforced activity, may have beneficial effects on the behavior of animals in captivity.

White (1975) showed large decreases in stereotypic rocking and social withdrawal when a pair of orangutans was moved from a relatively bare enclosure to one equipped with miscellaneous toys, climbable "jungle gyms," and other stimuli at the San Francisco Zoo. Even in cases where there is no apparent pathology, increasing the complexity of the stimulus environment can yield higher activity levels, which in turn provides more interesting displays of zoo visitors. Markowitz (1982), who had studied with Bennett and Rosenzweig, provides examples of this type of intervention with several different species, including polar bears (*Ursus arctos*), Asian elephants (*Elephantus maximus*), and Sumatran tigers (*Panthera tigris sumatrae*). More recently, Markowitz and Spinelli (1986) have begun an extensive program of behavioral enrichment in a facility housing captive primates used for several different types of behavioral and physiological research.

Enrichment of the environment for domestic livestock may also have beneficial effects on the animals' behavior. Animals such as pigs and calves are less aggressive and show other signs of improved well-being when raised in moderately complex situations (e.g., Wood-Gush, Stolba, and Miller, 1983). Environmental effects on the behavior of livestock are frequent topics in several publications, among them *Applied Animal Ethology* and *International Journal for the Study of Animal Problems*.

Environmental Enrichment as Treatment for the Effects of Undernutrition

Recent studies indicate that housing in an enriched environment can partially ameliorate the deleterious effects of undernutrition, although the details of these benefits remain subject to disagreement. Levitsky and Barnes (1972) found that the effects of early malnourishment (from birth through 7 weeks) were eliminated by housing in an enriched environment from 3 to 7 weeks (compared to handled or unhandled ICs) for open-field locomotion latency to enter a novel environment, aggressiveness (as measured by time spent fighting), and social following. Sara, King, and Lazarus (1976) reported, however, that malnourished IC rats were significantly impaired compared to SC performance on a spatial discrimination task, but that SC performance was equivalent to that of EC.

Katz and Davies (1983) provide evidence that previously malnourished rats show a response to environment identical in magnitude to normally-fed rats, leaving the malnourished but enriched rats with heavier cortices than malnourished, impoverished littermates. Bhide and Bedi obtained similar results for measures of forebrain weight, with EC exceeding IC, and normally fed rats exceeding malnourished rats, with no interaction (1982; 1984a). Environmental condition had a significant effect on numerical density, synaptic size, and estimates of synapse-to-neuron ratio (1984b), whereas there was no effect of undernutrition on these measures.

The only study known to the authors employing mice as subjects, however, Crnic (1983), using mice undernourished in infancy and subsequently housed in differential environments, reported finding no ameliorative effect of enrichment on brain weight or brain protein in spite of a higher level of performance on a test of passive avoidance and a higher level of activity in enriched mice. Whether this is a meaningful species difference or due to some methodological difference is at present unclear, and could only be resolved by parallel experiments using identical methodologies. Further work along these lines could provide useful information about how behavioral impairments caused by undernourishment or malnutrition in humans might be offset by experience.

Recovery from Brain Trauma

In the United States, approximately one person in 400 suffers from impairments related to head trauma. The social significance of the application of findings of environmentally-induced changes in the brain and behavior in this area would be enormous. The possibility that techniques developed in studies of differential environments might find practical application in aiding recovery from damage to any of several parts of the brain appears to be an area of research that is attracting major empirical attention in the 1980s. In view of the large numbers of studies in this area and the availability of a recent review of the effects of pre- and postoperative housing on behavioral sequelae to brain lesions (Dalrymple-Alford and Kelche, 1985), we will not attempt to provide encyclopedic coverage here. Instead, we will discuss a sampling of studies in the area.

When neonatal rats were subjected to cortical lesions, those subsequently housed in an enriched environment showed enhanced recovery of function (compared to littermates housed in IC) as measured by performance on the Hebb-Williams mazes (Schwartz, 1964). In later experiments, the beneficial effects of environmental enrichment was found to vary with severity of the lesion, in a manner reminiscent of Lashley's principle of mass action: the more cortex present, the larger the benefit of environmental enrichment (Will, Rosenzweig & Bennett, 1976). These beneficial effects were found not only after neonatal lesions but also when lesions were inflicted postweaning (Will, et al., 1977) or even in young adult rats (Will and Rosenzweig, 1976).

Whishaw, Zoborowski, and Kolb (1984) found that hemidecorticated rats were significantly improved in their performance of the Morris place-learning task (1984) by postsurgical enrichment when the lesions were inflicted in adulthood, but that enrichment had no effect on rats that had been hemidecorticated neonatally. Further, postsurgical environment had a significant effect on task performance, but not presurgical environment. This is not surprising when it is recalled that the clearest brain are found in neocortex, the tissue ablated in this technique.

In the rapidly developing field of neural tissue transplantation, recent evidence indicates that there may be an interaction between housing environment and the likelihood that tissue transplantation will lead to behavioral benefits (Will, Kelche, and Dalrymple-Alford, 1986): Subjects were given septal lesions, followed in some subjects by fetal septal grafts. Ten months later, deficits in Hebb-Williams maze performance were reduced in rats that had received transplanted tissue and housed postsurgically in enriched conditions, but not in those housed socially. Neither transplant nor enrichment was sufficient to produce behavioral recovery; applied together, significant behavioral effects were displayed. Two other laboratories, however, claim behavioral improvements following cell implantation without particular efforts to enrich the stimulus environment. Kesslak, Calin, Walencewicz, and Cotman (1986) found partial recovery of performance on an alternation task following cell implants at the site of hippocampal lesions, and Shapiro, Nilsson, Gage, Olton, and Bjorklund (1986) reported that implants improved performance on the Morris water maze spatial task after fimbria-fornix lesions. All of these reports are quite recent as of this writing; it remains to be seen whether procedural or lesion-site differences can account for these discrepant results.

The behavioral effects of hippocampal lesions can be partially reversed by experience in an enriched environment (Kelche and Will, 1978; Will and Kelche, 1979; Einon, Morgan, and Will, 1980), but there is evidence that neural responsiveness to the environment is reduced or eliminated by lesions in this structure as well (Kelche and Will, 1982).

Beneficial effects of enrichment on behavior are sometimes evident despite continued neural deterioration, both near the sites of cortical lesions and in remote loss of cortical tissue through secondary degeneration (Rosenzweig, Bennett, and Alberti, 1984). In animals with lesions to the entorhinal cortex, there is no apparent behavioral benefit from an enriched environment (Will, Kelche and Deluzarche, 1981), implying that certain neural machinery must be functional in order for the extra available environmental stimulation to be processed.

Davenport (1976) studied performance on numerous tasks of rats with experimentally-induced hypothyroidism and housed in impoverished, enriched or "superenriched" environments. Environmental enrichment led to significant reversals of hypothyroidism-induced deficits in performance compared to impoverished littermates on several tasks, including maze learning and retention, and barpressing extinction. Superenrichment led to at least partial behavioral recovery from a more severe hypothyroidism, while typical levels of environmental enrichment produced a level of recovery that was only inconsistently significant. Sjoden (1976) reported evidence from a complementary study of hyperthyroidism, wherein injections of thyroxine and tri-idothyronine were given in infancy, followed by 160 days in EC or IC. Spatial learning (T-maze) in adulthood was enhanced and the first reversal impaired by these injections for IC but not for EC rats.

A large proportion of the results indicating facilitation of performance in lesioned animals have used the Hebb-Williams mazes or similar problems (see Dalrymple-Alford and Kelche, 1985, for an extensive discussion). Consistent with the behavioral differences described above in Chapter 3, several experiments show an absence of enrichment-induced amelioration of lesion effects for relatively simpler tasks. The hypothyroid EC and IC groups of Davenport (1976) did not differ on acquisition of barpressing or passive avoidance; enrichment does not aid rats with septal lesions on performance of alternation tasks (Donovick, Burright, and Swidler, 1973); nor does enrichment facilitate avoidance performance in rats with serial or simultaneous septal-forebrain lesions (Fass, Wrege, Greenough, and Stein, 1980). Stein, Finger, and Hart (1983) have suggested that enrichment is most likely to ameliorate lesion-induced performance deficits when the task allows substitution of any one of several different information channels (e.g., visual, kinesthetic, or tactile) for completion, and least likely to be beneficial on more specific tasks. Supporting evidence is provided by Cornwell and Overman (1981), demonstrating that enrichment facilitates Hebb-Williams maze learning, but not visual discrimination, in kittens given neonatal lesions to the visual cortex.

Held, Gordon, and Gentile (1985) reported that performance of a complex motor task is facilitated by enrichment in rats having lesions to the motor cortex. A microanalysis of subjects' movement patterns revealed that rats preoperatively enriched and impoverished showed postoperative differences in movement topology while performing the motor task that were only partially diminished by postoperative environment. The placement of the lesion at the site of cortical control of the task used in this study, combined with the differential effect of the lesion, leads to the inference that the EC subjects had more highly distributed control of performance of this task than IC.

Several studies have reported that relatively brief experience in enriched environments produces effects that are in many respects comparable to those produced by 24-hour enrichment (e.g., Rosenzweig, Love, and Bennett, 1968; Ferchmin and Eterovic, 1986). At least two studies have found that two hours per day of environmental enrichment produced as much behavioral improvement in rats with neonatally-induced cortical and hippocampal lesions as did 24 hours per day (Will, Rosenzweig, Bennett, Hebert and Morimoto, 1977; Will, Sutter, and Offerlin, 1977). A third (Held, et al., 1985, discussed above) reports significant ameliorative effect with two hours per day in EC, without making comparisons to 24-hour EC subjects. The probability that practical human applications for rehabilitation can be developed is substantially increased by the finding that enrichment during a limited time can enhance recovery from brain trauma.

A few books concerned with applications have already called attention to earlier stages of this research: Walsh and Greenough (1976) and Bach-y-Rita (1980).

Aging

The findings of continued plasticity in the brains and behavior of older animals have led to investigations concerning the possible therapeutic role of environmental enrichment in behavioral function in old age. Riege (1971) found that plasticity of the brain persisted well into adulthood in the rat, and Greenough, et al. (1986) report that housing in EC can moderate cerebellar degeneration and prompt new dendrite growth in two-year-old rats. Doty (1972) found, in rats differentially housed from 300 to 660 days of age, that initial avoidance learning did not differentiate the EC and IC groups, but EC were superior on reversal tests following this learning; this difference occurred in the absence of differences in pain threshold for the footshock used to motivate the avoidance task. Kubanis, Zornetzer, and Freund (1982) found that enrichment improved performance on passive avoidance tasks in adult and aging mice, even in the absence of clear amelioration of age-related decline of binding in cholinergic receptors.

These findings have inspired attempts to improve the well-being and fitness of elderly persons. One such attempt at improving the environment for the aged, with direct antecedents in laboratory enrichment studies, is a "foster grandparent" program in Southern California (Sandman and Donnelly, 1983). In a review of life-span development, Honzik (1984) discussed whether there are genuine and/or necessary declines in ability with age and pointed out "...the measurable effects of stimulation on the brains of older rats ... should be kept in mind" (p. 315).

Potential for Additional Applications

The effects of differential environments on the brain are also being used as either an independent or a dependent variable in many different types of endeavor. For example, the 1986 volume of *Society for Neuroscience Abstracts* contains numerous reports of studies from 10 different laboratories using environmental differences as a methodological factor in studies of diverse neurobiological phenomena, approximately evenly split between studies primarily intended to provide more information concerning the brain's response to manipulation of environmental complexity and studies making use of known EC-IC effects in the study of other phenomena.

The interaction of environmentally-induced changes in neural and behavioral variables with factors such as genetics, brain trauma, or the effects of toxins has also attracted attention (e.g., Will, et al., 1986). Based on his observation of complex genotype-environment interactions on brain weights in mice, Henderson (1973) argued that behavior genetic research in the laboratory is in danger of drawing erroneous conclusions if the standard (i.e., in most cases, relatively impoverished) laboratory housing environment is used exclusively in behavior genetics. Another report, after providing evidence that drug and environmental effects are interactive, argued for the use of both enriched and impoverished environments in all studies of developmental toxicology: "the consequences of perinatal drug treatments ... may well go undetected unless the response of the drug-treated rats to differential rearing is measured" (Pappas, et al., 1984).

It seems quite likely that in the future the results of this type of studies will find many additional applications, aiding the human condition both directly and indirectly. The possibility that manipulation of the external environment can have measurable effects on functionally significant aspects of human brain function and/or behavior makes the understanding of these phenomena critically important.

While the neural processes that serve as the substrate for complex problem-solving and intelligence are still largely inscrutable, impressive advances are being made towards an understanding of the memory systems that serve cognitive and habit-based behaviors (Mishkin, Malamut and Bachevalier, 1984; Squire & Butters, 1984; Matthies, 1986). Progress is also being made in understanding the specific neural changes that underlie the storage of memory for conditioning and simple tasks (for multiple approaches to this question see, for example, the review volumes edited by Lynch, McGaugh, and Weinberger, 1984; Squire & Butters, 1984; Weinberger, McGaugh, and Lynch, 1985). The body of experimental evidence discussed here could provide insights into the neural underpinnings of these processes (and possible beneficial applications thereof) but only when causes for these changes are better understood.

Conclusion

The central empirical finding of what is known to many as the "EC-IC literature," that differences in behavioral experience not profoundly different from what an animal might experience in the natural world can produce substantial changes in the structure and function of the nervous system, was reported a quarter of a century ago. Although general interest in the problem has waxed and waned in the intervening time, the time of this writing is one of widespread empirical efforts toward a comprehensive understanding of behaviorally induced neural plasticity. In addition, attention is once again being focused upon the behavioral factors involved in this complex phenomenon, both of the type resulting from nervous system changes and those that may be causally implicated in those changes. It is clear that what goes on in the different environments is one of the keys to unravelling the differences created by those environments, and is a problem in the best psychobiological tradition of deciphering brain-behavior relationships.

Even as the empirical searches for evidence concerning cross-species generality, and neuroanatomical, neurochemical, and behavioral sequelae continue, a second approach may illuminate these phenomena further: If the causes of the brain and behavioral changes resulting from enriched experience were shown to be due to psychobiological processes that are common across species, then the inference of cross-species generality and therefore the applicability to human problems would not rest solely on the empirical evidence provided to date but also upon sound theoretical foundations.

The ability of the individual organism to alter its own neural structure in response to environmental demands is at the heart of the biological processes underlying behavioral plasticity, of which learning and memory are a part. An understanding of the causes and behavioral consequences of this ability is a critical element in a solution to the problem of the nature of adaptive behavior.

References

Aitken, P. P. (1974). Early experience, emotionality, and exploration in the rat: a critique of Whimbey and Denenberg's Hypothesis. *Developmental Psychobiology*, 7, 129-134.

Alkon, D. L. (1985). Conditioning-induced changes of *Hermissenda* channels: relevance to mammalian brain function. In N. M. Weinberger, J. L. McGaugh, and G. Lynch (Eds.), *Memory Systems of the Brain*, 9-26. New York: Guilford Press.

Altman, J. (1969). Autoradiographic and histological studies of postnatal neurogenesis. *Journal of Comparative Neurology*, **136**, 269-294.

Altman, J., Wallace, R. B., Anderson, W. J., & Das, G. D. (1968). Behaviorally induced changes in length of cerebrum in rats. *Developmental Psychobiology*, 1, 112-117.

Altschuler, R. A. (1979). The effects of increased experience and training of synaptic density in Area CA3 of the rat hippocampus. *Dissertation Abstracts International*, **39(9B)**, 4140-4141.

Aulich, D. (1976). Escape versus exploratory activity: an interpretation of rats' behavior in the open field and a light-dark preference test. *Behavioral Processes*, 1, 153-164.

Bach-y-Rita, P. (1980). *Recovery of Function: Theoretical Considerations for Brain Injury Rehabilitation*. Bern, Switzerland: Hans Huber.

Baenninger, L. P. (1967). Comparison of behavioural development in socially isolated and grouped rats. *Animal Behaviour*, **15**, 312-323.

Beaulieu, C., & Colonnier, M. (1985). The effects of environmental complexity on the numerical density of neurons and on the size of their nuclei in the visual cortex of cat. *Society for Neuroscience Abstracts*, **11**, 225.

Beaulieu, C., & Colonnier, M. (1986). The effects of impoverished and enriched environments on the number and size of boutons containing flat vesicles in the visual cortex of cat. *Society for Neuroscience Abstracts*, **12**, 128.

Benloucif, S., Rosenzweig, M. R., & Bennett, E. L. (1984). The effect of norepinephrine depletion by xylamine on investigatory behavior and on brain weights with enriched rearing. *Society for Neuroscience Abstracts*, **10**, 1174.

Bennett, E. L. (1976). Cerebral effects of differential experience and training. In: M. R. Rosenzweig and E. L. Bennett (Eds.), *Neural Mechanisms of Learning and Memory*, 279-289. Cambridge: The MIT Press.

Bennett, E. L., Diamond, M. C., Krech, D., & Rosenzweig, M. R. (1964a). Chemical and anatomical plasticity of brain. *Science*, **146**, 610-619.

Bennett, E. L., Diamond, M. C., Krech, D., & Rosenzweig, M. R. (1969). Rat brain: Effect of environmental enrichment on wet and dry weights. *Science*, **163**, 825-826.

Bennett, E. L., Krech, D., & Rosenzweig, M. R. (1964b). Reliability and specificity of cerebral effects of environmental complexity and training. *Journal of Comparative and Physiological Psychology*, **57**, 440-441.

Bennett, E. L., Orme, A., & Hebert, M. (1972). Cerebral protein synthesis inhibition and amnesia produced by scopolamine, cycloheximide, streptovitacin A, anisomycin, and emetine in rat. *Federation Proceedings*, **31**, 838.

Bennett, E. L., & Rosenzweig, M. R. (1981). Behavioral and biochemical methods to study brain responses to environment and experience. In R. Lahue (Ed.), *Methods in Neurobiology*, volume 2, 101-141. New York: Plenum Press.

Bennett, E. L., Rosenzweig, M. R., & Diamond, M. C. (1970). Time courses of effects of differential experience on brain measures and behavior of rats. In W. L. Byrne (Ed.), *Molecular Approaches to Learning and Memory*, 55-89. New York: Academic Press.

Bennett, E. L., Rosenzweig, M. R., Diamond, M. C., Morimoto, H., & Hebert, H. (1974). Effects of successive environments on brain measures. *Physiology and Behavior*, **12**, 621-631.

Bennett, E. L., Rosenzweig, M. R., Morimoto, H., & Hebert, M. (1979). Maze training alters brain weights and cortical RNA/DNA ratios. *Behavioral and Neural Biology*, **26**, 1-22.

Bernstein, L. (1973). A study of some enriching variables in a free environment for rats. *Journal of Psychosomatic Research*, **17**, 85-88.

Bhide, P. G., & Bedi, K. S. (1982). The effects of environmental diversity on well-fed and previously undernourished rats: I. Body and brain measurements. *The Journal of Comparative Neurology*, **207**, 403-409.

Bhide, P. G., & Bedi, K. S. (1984a). The effects of a lengthy period of environmental diversity on well-fed and previously undernourished rats. I. Neurons and glial cells. *Journal of Comparative Neurology*, **227**, 296-304.

Bhide, P. G., & Bedi, K. S. (1984b). The effects of a lengthy period of environmental diversity of well-fed and previously undernourished rats. II. Synapse-to-neuron ratios. *Journal of Comparative Neurology*, **227**, 305-310.

Bingham, W. E., & Griffiths, W. J. (1952). The effects of different environments during infancy on adult behavior in the rat. *Journal of Comparative and Physiological Psychology*, **45**, 307-312.

Black, J. E., Parnisari, R., Eichbaum, E., & Greenough, W. T. (1986). Morphological effects of housing environment and voluntary exercise on cerebral cortex and cerebellum of old rats. *Society for Neuroscience Abstracts*, **12**, 1579.

Brown, C. P. (1971). Cholinergic activity in rats following enriched stimulation and training: direction and duration of effects. *Journal of Comparative and Physiological Psychology*, **75**, 408-416.

Brown, C. P., & King, M. G. (1971). Developmental environment: variables important for later learning and changes in cholinergic activity. *Developmental Psychobiology*, **4**, 275-286.

Brown, R. T. (1968). Early experience and problem-solving ability. *Journal of Comparative and Physiological Psychology*, **65**, 433-440.

Caul, W. F., Freeman, B. J., & Buchanan, D. C. (1974). Effects of differential rearing condition on heart rate conditioning and response suppression. *Developmental Psychobiology*, **8**, 63-68.

Chang, F.-L. F., & Greenough, W. T. (1978). Increased dendritic branching in hemispheres opposite eyes exposed to maze training in split-brain rats. *Society for Neuroscience Abstracts*, **4**, 469.

Chang, F.-L. F., & Greenough, W. T. (1982). Lateralized effects of monocular training on dendritic branching in adult split-brain rats. *Brain Research*, **232**, 283-292.

Chang, S.-Y. (1969). *Changes produced in the rat brain by environmental complexity and drug injection*. Unpublished doctoral dissertation, University of California, Berkeley.

Cheal, M., Foley, K., & Kastenbaum, R. (1984). Enrichment for one hour a month resulted in accelerated growth in adult gerbils. *Society for Neuroscience Abstracts*, **10**, 452.

Cheal, M., Foley, K., & Kastenbaum, R. (1986). Brief periods of environmental enrichment facilitate adolescent development of gerbils. *Physiology and Behavior*, **36**, 1047-1051.

Collins, R. A. (1970). Experimental modification of brain weight and behavior in mice: an enrichment study. *Developmental Psychobiology*, **3**, 145-155.

Colonnier, M., & Beaulieu, C. (1985). The differential effect of impoverished and enriched environments on the number of "round asymmetrical" and "flat symmetrical" synapses in the visual cortex of cat. *Society for Neuroscience Abstracts*, **11**, 226.

Connor, J. R., Melone, J. H., Yuen, A. R., & Diamond, M. C. (1981). Dendritic length in aged rats' occipital cortex: An environmentally induced response. *Experimental Neurology*, **73**, 827-830.

Cooper, R. M., & Zubeck, J. P. (1958). Effects of enriched and restricted early environments on the learning ability of bright and dull rat. *Canadian Journal of Psychology*, **12**, 159-164.

Cordoba, F., Yusta, B., & Munoz-Blanco, J. (1984). Changes in neurotransmitter amino acids and protein in CNS areas of mice subjected to differential housing conditions. *Pharmacology, Biochemistry and Behavior*, **21**, 349-352.

Cornwell, P., & Overman, W. (1981). Behavioral effects of early rearing conditions and neonatal lesions on the visual cortex in kittens. *Journal of Comparative and Physiological Psychology*, **95**, 848-862.

Coyle, I. R., & Singer, G. (1975a). The interaction of post-weaning housing conditions and prenatal drug effects of behavior. *Psychopharmocologia*, **41**, 237-244.

Coyle, I. R., & Singer, G. (1975b). The interactive effects of prenatal imipramine exposure and postnatal rearing conditions on behaviour and histology. *Psychopharmacologia*, **44**, 253-256.

Crnic, L. (1983). Effects of nutrition and environment on brain biochemistry and behavior. *Developmental Psychobiology*, **16**, 129-145.

Cummins, R. A., Livesey, P. J., & Bell, J. A. (1982). Cortical depth changes in enriched and isolated mice. *Developmental Psychobiology*, **15**, 187-195.

Cummins, R. A., Livesey, P. J., & Bell, J. A. (1983). Erratum: Cortical depth changes in enriched and isolated mice. *Developmental Psychobiology*, **16**, 159.

Cummins, R. A., Livesey, P. J., Evans, J. G. M., & Walsh, R. N. (1977). A developmental theory of environmental enrichment. *Science*, **197**, 692-694.

Cummins, R. A., Walsh, R. N., Budtz-Olsen, O. E., Konstantinos, T., & Horsfall, C. R. (1973). Environmentally-induced changes in the brains of elderly rats. *Nature*, **243**, 516-518.

Dalrymple-Alford, J. C., Benton, D., & Brain, P. F. (1983). *Behavourial comparisons of differentially housed rats: Enriched housing, group housing, and social isolation*. Unpublished manuscript.

Dalrymple-Alford, J. C., & Kelche, C. B. (1985). Behavioural effects of preoperative and postoperative differential housing in rats with brain lesions: A review. In B. E. Will, P. Schmitt, and J. C. Dalrymple-Alford (Eds.), *Brain Plasticity, Learning, and Memory*, 441-458. New York: Plenum Press.

Darwin, C. (1859). *On the Origin of Species*. London: John Murray.

Davenport, J. W. (1976). Environmental therapy in hypothyroid and other disadvantaged animal populations. In R. N. Walsh and W. T. Greenough (Eds.), *Environments as Therapy for Brain Dysfunction*, 71-114. New York: Plenum Press.

Dell, P. A., & Rose, F. D. (1986). The impairing effects on environmental impoverishment in rats: a cognitive deficit?. *IRCS Medical Sciences*, **14**, 19-20.

Denenberg, V. H. (1969). Open-field behavior in the rat: what does it mean?. *Annals New York Academy of Sciences*, **159**, 852-859.

Denenberg, V. H., & Morton, J. R. C. (1962a). Effects of environmental complexity and social groupings upon modification of emotional behavior. *Journal of Comparative and Physiological Psychology*, **55**, 242-246.

Denenberg, V. H., & Morton, J. R. C. (1962b). Effects of preweaning and postweaning manipulations upon problem-solving behavior. *Journal of Comparative and Physiological Psychology*, **55**, 1096-1098.

Denenberg, V. H., Woodcock, J. M., & Rosenberg, K. M. (1968). Long-term effects of preweaning and postweaning free-environment experience on rats problem-solving behavior. *Journal of Comparative and Physiological Psychology*, **66**, 533-535.

Devenport, L., Dallas, S., Carpenter, C., & Renner, M. J. (1986). The relationship between enrichment- and adrenalectomy-induced cortical growth. *Society for Neuroscience Abstracts*, **12**, 1222.

Diamond, M. C. (1976). Anatomical brain changes induced by environment. In L. Petrinovich and J. L. McGaugh (Eds.), *Knowing, Thinking and Believing*, 215-241. New York: Plenum Press.

Diamond, M. C., Johnson, R. E., & Ingham, C. A. (1971). Brain plasticity induced by environment and pregnancy. *International Journal of Neuroscience*, **2**, 171-178.

Diamond, M. C., Krech, D., & Rosenzweig, M. R. (1964). The effects of an enriched environment of the histology of the rat cerebral cortex. *Journal of Comparative Neurology*, **123**, 111-120.

Diamond, M. C., Law, F., Rhodes, H., Lindner, B., Rosenzweig, M. R., Krech, D., & Bennett, E. L. (1966). Increases in cortical depth and glia numbers in rats subjected to enriched environments. *Journal of Comparative Neurology*, **128**, 117-126.

Diamond, M. C., Lindner, B., Johnson, R., Bennett, E. L., & Rosenzweig, M. R. (1975). Differences in occipital cortical synapses from environmentally enriched, impoverished, and standard colony rats. *Journal of Neuroscience Research*, **1**, 109-119.

Diamond, M. C., Lindner, B., & Raymond, A. (1967). Extensive cortical depth measurements and neuron size increases in the cortex of environmentally enriched rats. *Journal of Comparative Neurology*, **131**, 357-364.

Diamond, M. C., Rosenzweig, M. R., & Krech, D. (1965). Relationships between body weight and skull development in rats raised in enriched and impoverished conditions. *Journal of Experimental Zoology*, **160**, 29-36.

Domjan, M., Schorr, R., & Best, M. (1977). Early environmental influences on conditioned and unconditioned ingestional and locomotor behavior. *Developmental Psychobiology*, **10**, 499-506.

Donovick, P. J., Burright, R. G., & Swidler, M. A. (1973). Presurgical rearing environment alters exploration, fluid consumption and learning of septal lesioned and control rats. *Physiology and Behavior*, **11**, 543-553.

Doty, B. A. (1972). The effect of cage environment upon avoidance responding of aged rats. *Journal of Gerontology*, **27**, 358-360.

Dyson, S. E., & Jones, D. G. (1980). Quantitation of terminal parameters and their interrelationships in maturing central synapses: a perspective for experimental studies. *Brain Research*, **183**, 43-59.

Edwards, H. P., Barry, W. F., & Wyspianski, J. O. (1969). Effects of differential rearing on photic evoked potentials and brightness discrimination in the albino rat. *Developmental Psychobiology*, **2(3)**, 133-138.

Einon, D. F., Humphreys, A. P., Chivers, S. M., Field, S., & Naylor, V. (1981). Isolation has permanent effects upon the behavior of the rat, but not the mouse, gerbil, or Guinea pig. *Developmental Psychobiology*, **14**, 343-355.

Einon, D. F., & Morgan, M. J. (1977). A critical period for social isolation in the rat. *Developmental Psychobiology*, **10**, 123-132.

Einon, D. F., Morgan, M. J., & Kibbler, C. C. (1978). Brief periods of socialization and later behavior in the rat. *Developmental Psychobiology*, **11**, 213-225.

Einon, D. F., Morgan, M. J., & Will, B. E. (1980). Effects of post-operative environment on recovery from dorsal hippocampal lesions in young rats: Tests of spatial memory and motor transfer. *Quarterly Journal of Experimental Psychology*, **32**, 137-148.

Eterovic, V. A., & Ferchmin, P. A. (1986). A decrease in putrescine enhances brain plasticity elicited by complex environment. *Society for Neuroscience Abstracts*, **12**, 1167.

Fagen, R. (1981). *Animal Play Behavior*. Oxford: Oxford University Press.

Fagen, R. (1982). Evolutionary issues in the development of behavioral flexibility. In P. P. G. Bateson and R. A. Hinde (Eds.), *Perspectives in Ethology*, Volume 5, 364-383. New York: Plenum Press.

Fass, B., Wrege, K., Greenough, W. T., & Stein, D. G. (1980). Behavioral symptoms following serial or simultaneous septal-forebrain lesions: similar syndromes. *Physiology and Behavior*, **25**, 683-690.

Ferchmin, P. A., Bennett, E. L., & Rosenzweig, M. R. (1975). Direct contact with enriched environment is required to alter cerebral weights in rats. *Journal of Comparative and Physiological Psychology*, **88**, 360-367.

Ferchmin, P. A., & Eterovic, V. A. (1986). Forty minutes of experience increase the weight and RNA content of cerebral cortex in periadolescent rats. *Developmental Psychobiology*, **19**, 511-519.

Ferchmin, P. A., & Eterovic, V. A. (1987). Role of polyamines in experience-dependent brain plasticity. *Pharmacology, Biochemistry, and Behavior*, in press.

Ferchmin, P. A., Eterovic, V. A., & Caputto, R. (1970). Studies of brain weight and RNA content after short periods of exposure to environmental complexity. *Brain Research*, **20**, 49-57.

Ferchmin, P. A., Eterovic, V. A., & Levin, L. E. (1980). Genetic learning deficiency does not hinder environment-dependent brain growth. *Physiology and Behavior*, **24**, 45-50.

Ferrer, I. (1983). Cambios morfologicos en la corteza cerebral de retones somtidos a medios enriquecidos y a medios empobrecidos en estimulos sensoriales y su posterior recuperacion. *Archivos de Neurobiologia*, **46(3)**, 177-182 (English Abstract).

Fiala, B. A., Joyce, J. N., & Greenough, W. T. (1978). Environmental complexity modulates growth of granule cell dendrites in developing but not adult hippocampus of rats. *Experimental Neurology*, **59**, 372-383.

Fiala, B., Snow, F. M., & Greenough, W. T. (1977). "Impoverished" rats weigh more than "enriched" rats because they eat more. *Developmental Psychobiology*, **10**, 537-541.

Flicker, C., & Geyer, M. A. (1982). The hippocampus as a possible site of action for increased locomotion during intracerebral infusions on norepinephrine. *Behavioral and Neural Biology*, **34**, 421-426.

Floeter, M. K., & Greenough, W. T. (1978). Cerebellar plasticity: Modification of dendritic branching by differential rearing in monkeys. *Society for Neuroscience Abstracts*, **4**, 471.

Floeter, M. K., & Greenough W. T. (1979). Cerebellar plasticity: modification of Purkinje cell structure by differential rearing in monkeys. *Science*, **206**, 227-229.

Flood, J. F., Rosenzweig, M. R., Bennett, E. L., & Orme, H. E. (1973). The influence of duration of protein synthesis inhibition on memory. *Physiology and Behavior*, **10**, 555-562.

Forgays, D. G., & Forgays, J. W. (1952). The nature of the effect of free-environment experience in the rat. *Journal of Comparative and Physiological Psychology*, **45**, 322-328.

Freeman, B. J., & Ray, O. S. (1972). Strain, sex, and environmental effects on appetitively and aversively motivated learning tasks. *Developmental Psychobiology*, **5**, 101-109.

Gates, E. (1909). Biographical sketch. *The National Cyclopedia of American Biography*, Volume X, 354. New York: James T. White & Co.

Geller, E., Yuwiler, A., & Zolman, J. F. (1965). Effects of environmental complexity on constituents of brain and liver. *Journal of Neurochemistry*, **12**, 949-955.

Gibbs, M., & Ng, K. T. (1977). Psychobiology of memory: Towards a model of memory formation. *Biobehavioral Reviews*, 1, 113-136.

Gill, J. H., Reid, L. D., & Porter, P. B. (1966). Effects of restricted rearing on Lashley-stand performance. *Psychological Reports*, **19**, 239-242.

Globus, A., Rosenzweig, M. R., Bennett, E. L., & Diamond, M. C. (1973). Effects of differential environments on dendritic spine counts. *Journal of Comparative and Physiological Psychology*, **82**, 175-181.

Gluck, J. P., Harlow, H. F., & Schiltz, K. A. (1973). Differential effects of early enrichments and deprivation on learning in the Rhesus monkey (Macaca mulatta). *Journal of Comparative and Physiological Psychology*, **84**, 598-604.

Goodlett, C. R., Engellenner, W. J., Burright, R. G., & Donovick, P. J. (1982). Influence of environmental rearing history and postsurgical environmental change on the septal rage syndrome in mice. *Physiology and Behavior*, **28**, 10771081.

Green, E. J., & Greenough, W. T. (1986). Altered synaptic transmission in dentate gyrus of rats reared in complex environments: evidence from hippocampal slices maintained in vitro. *Journal of Neurophysiology*, **55**, 739-749.

Green, E. J., Greenough, W. T., & Schlumpf, B. E. (1983). Effects of complex or isolated environments on cortical dendrites of middle-aged rats. *Brain Research*, **264**, 233-240.

Greenough, W. T. (1976). Enduring brain effects of differential experience and training. *Neural Mechanisms of Learning and Memory*, 255-278. Cambridge: M.I.T. Press.

Greenough, W. T., Hwang, H.-M. F., & Corman, C. (1985). Evidence for active synapse formation, or altered postsynaptic metabolism, in visual cortex of rats reared in complex environments. *Proceedings of the National Academy of Sciences*, **82**, 4549-4552.

Greenough, W. T., Juraska, J. M., & Volkmar, F. R. (1979). Maze training effects on dendritic branching in occipital cortex of adult rats. *Behavioral and Neural Biology*, **26**, 287-297.

Greenough, W. T., Larson, J. R., & Withers, G. S. (1985). Effects of unilateral and bilateral training in a reaching task on dendritic branching of neurons in the rat motor-sensory forelimb cortex. *Behavioral and Neural Biology*, **44**, 301-313.

Greenough, W. T., Madden, T. C., & Fleischmann, T. B. (1972). Effects of isolation, daily handling, and enriched rearing on maze learning. *Psychonomic Science*, **27**, 279-280.

Greenough, W. T., McDonald, J. W., Parnisari, R. M., & Camel, J. E. (1986). Environmental conditions modulate degeneration and new dendrite growth in cerebellum of senescent rats. *Brain Research*, **380**, 136-143.

Greenough, W. T., & Volkmar. F. R. (1973). Pattern of dendritic branching in occipital cortex of rats reared in complex environments. *Experimental Neurology*, 40, 491-504.

Greenough, W. T., Volkmar, F. R., & Juraska, J. M. (1973). Effects of rearing complexity on dendritic branching in frontolateral and temporal cortex of the rat. *Experimental Neurology*, **41**, 371-378.

Greenough, W. T., West, R. W., & DeVoogd, T. J. (1978). Subsynaptic plate perforations: Changes with age and experience in the rat. *Science*, 202, 1096-1098.

Greenough, W. T., Wood, W. E., & Madden, T. C. (1972). Possible memory storage differences among mice reared in environments varying in complexity. *Behavioral Biology*, **7**, 717-722.

Greenough, W. T., Yuwiler, A., & Dollinger, M. (1973). Effects of posttrial eserine administration of learning in "enriched"-and "impoverished"-reared rats. *Behavioral Biology*, **8**, 261-272.

Greer, E. R., Diamond, M. C., & Murphy, G. M. (1982). Increased branching of basal dendrites on pyramidal neurons in the occipital cortex of homozygous Brattleboro rats in standard and enriched environmental conditions: a Golgi study. *Experimental Neurology*, **76**, 254-262.

Grouse, L. D., Schrier, B. K., Bennett, E. L., Rosenzweig, M. R., & Nelson, P. G. (1978). Sequence diversity studies of rat brain RNA: Effects of environmental complexity on rat brain RNA diversity. *Journal of Neurochemistry*, **30**, 191-203.

Gutwein, B. M., & Fishbein, W. (1980a). Paradoxical sleep and memory (I): Selective alterations following enriched and impoverished environmental rearing. *Brain Research Bulletin*, **5**, 9-12.

Gutwein, B. M., & Fishbein, W. (1980b). Paradoxical sleep and memory (II): Sleep circadian rhythmicity following enriched and impoverished environmental rearing. *Brain Research Bulletin*, **5**, 105-109.

Hamilton, W. L., Diamond, M. C., Johnson, R. E., & Ingham, C. A. (1977). Effects of pregnancy and differential environments on rat cerebral cortical depth. *Experimental Neurology*, **19**, 333-340.

Hawkins, R. B., & Kandel, E. R. (1984). Steps toward a cell-biological alphabet for elementary forms of learning. In G. Lynch, J. L. McGaugh, and N. M. Weinberger (Eds.), *Neurobiology of Learning and Memory*, 385-404.

Hebb, D. O. (1947). The effects of early experience on problem-solving at maturity. *American Psychologist*, **2**, 306-307.

Hebb, D. O., & Williams, K. (1946). A method of rating animal intelligence. *Journal of General Psychology*, **34**, 56-65.

Held, J. M., Gordon, J., & Gentile, A. M. (1985). Environmental influences on locomotor recovery following cortical lesions in rats. *Behavioral Neuroscience*, **99**, 678-690.

Henderson, N. D. (1970). Brain weight increases resulting from environmental enrichment: a directional dominance in mice. *Science*, **169**, 776-778.

Henderson, N. D. (1973). Brain weight changes resulting from enriched rearing conditions: a diallele analysis. *Developmental Psychobiology*, **6**, 367-376.

Henderson, N. D. (1979). Genetic correlations between brain size and some behaviors of housemice. In M. E. Hahn, C. Jensen, and B. C. Dudek (Eds.) *Development of Evolution of Brain Size*, 347-370. New York: Academic Press.

Holloway, R. L. (1966). Dendritic branching in the rat visual cortex. Effects of extra environmental complexity and training. *Brain Research*, **2**, 393.

Holson, R. R. (1986). Feeding neophobia: a possible explanation for the differential maze performance of rats reared in enriched or isolated environments. *Physiology and Behavior*, **38**, 191-201.

Honzik, M. (1984). Life-span development. *Annual Review of Psychology*, **35**, 309-331.

Hughes, K. R. (1965). Dorsal and ventral hippocampus lesions and maze learning: Influences of preoperative environment. *Canadian Journal of Psychology*, **19**, 325-332.

Hymovitch, B. (1952). The effects of experimental variations on problem-solving in the rat. *Journal of Comparative and Physiological Psychology*, **45**, 313-321.

Johnston, T. D. (1982). Selective costs and benefits in the evolution of learning. *Advances in the Study of Behavior*. New York: Academic Press.

Jones, D. G., & Smith, B. J. (1980a). The hippocampus and its response to differential environments. *Progress in Neurobiology*, **15**, 19-69.

Jones, D. G., & Smith, B. J. (1980b). Morphological analysis of the hippocampus following differential rearing in environments of varying social and physical complexity. *Behavioral and Neural Biology*, **30**, 135-147.

Jorgensen, O. S., & Meier, E. (1979). Microtubular proteins in the occipital cortex of rats housed in enriched and in impoverished environments. *Journal of Neurochemistry*, **33**, 381-382.

Joseph, R. (1979). Effects of rearing and sex on maze learning and competitive exploration in rats. *The Journal of Psychology*, **101**, 37-43.

Juraska, J. M. (1984). Sex differences in developmental plasticity in the visual cortex and hippocampal dentate gyrus. *Sex Differences in the Brain. Progress in Brain Research*, **61**, 205-214. Amsterdam: Elsevier/North-Holland.

Juraska, J. M. (1984). Sex differences in dendritic response to differential experience in the rat visual cortex. *Brain Research*, **295**, 27-34.

Juraska, J. M., Fitch, J. M., Henderson, C., & Rivers, N. (1985). Sex differences in the dendritic branching of dentate granule cells following differential experience. *Brain Research*, **333**, 73-80.

Juraska, J. M., Greenough, W. T., & Conlee, J. W. (1983). Differential rearing affects responsiveness of rats to depressant and convulsant drugs. *Physiology and Behavior*, **31**, 711-715.

Juraska, J. M., Greenough, W. T., Elliot C., Mack, K. J., & Berkowitz, R. (1980). Plasticity in adult rat visual cortex: An examination of several cell populations after differential rearing. *Behavioral and Neural Biology*, **29**, 157-167.

Juraska, J. M., Henderson, C., & Muller, J. (1984). Differential rearing experience, gender, and radial maze performance. *Developmental Psychobiology*, **17**, 209-215.

Juraska, J. M., & Meyer, M. (1986). Behavioral interactions of postweaning male and female rats with a complex environment. *Developmental Psychobiology*, **19**, 493-500.

Kasamatsu, T., Pettigrew, J. D., & Ary, M. (1981). Cortical recovery from effects of monocular deprivation. Acceleration with norepinephrine and suppression with 6-hydroxydopamine. *Journal of Neurophysiology*, **45**, 254-266.

Katz, H. B., & Davies, C. A. (1983). The separate and combined effects of early undernutrition and environmental complexity at different ages on cerebral measures in rats. *Developmental Psychobiology*, **16**, 47-58.

Kelche, C. R., & Will, B. (1978). Effects of environment on functional recovery after hippocampal lesions in adult rats. *Physiology and Behavior*, **21**, 935-941.

Kelche, C. R. & Will, B. (1982). Effects of postoperative environments following dorsal hippocampal lesions on dendritic branching and spines in rat occipital cortex. *Brain Research*, **245**, 107-115.

Kesslak, J. P., Calin, L., Walencewicz, A., & Cotman, C. W. (1986). Adult brain transplants facilitate behavioral recovery. *Society for Neuroscience Abstracts*, **12**, 1473.

Kety, S. S. (1970). The biogenic amines in the central nervous system: Their possible roles in arousal, emotion, and learning. *The Neurosciences: Second Study Program*, 324-336. New York: Rockefeller University Press.

Kiyono, S., Seo, M. L., & Shibagahi, M. (1981). Effects of rearing environments upon sleep-waking parameters in rats. *Physiology and Behavior*, **26**, 391-395.

Kopcik, J. R., Juraska, J. M., & Washburne, D. L. (1986). Sex and environmental effects on the ultrastructure of the rat corpus callosum. *Society for Neuroscience Abstracts*, **12**, 1218.

Krech, D., Rosenzweig, M. R., & Bennett, E. L. (1956). Dimensions of discrimination and level of cholinesterase activity in the cerebral cortex of the rat. *Journal of Comparative and Physiological Psychology*, **49**, 261-268.

Krech, D., Rosenzweig, M. R., & Bennett, E. L. (1960). Effects of environmental complexity and training on brain chemistry. *Journal of Comparative and Physiological Psychology*, **53**, 509-519.

Krech, D. Rosenzweig, M. R., & Bennett, E. L. (1962). Relations between brain chemistry and problem-solving among rats raised in enriched and impoverished environments. *Journal of Comparative and Physiological Psychology*, **55**, 801-807.

Krech, D., Rosenzweig, M. R., & Bennett, E. L. (1966). Environmental impoverishment, social isolation and changes in brain chemistry and anatomy. *Physiology and Behavior*, **1**, 99-104.

Kubanis, P., Zornetzer, S. F., & Freund, G. (1982). Memory and postsynaptic cholinergic receptors in aging mice. *Pharmacology, Biochemistry and Behavior*, **17**, 313-322.

Kuenzle, C. C., & Knusel, A. (1974). Mass training of rats in a superenriched environment. *Physiology and Behavior*, **13**, 205-210.

La Torre, J. C. (1968). Effect of differential environmental enrichment on brain weight and on acetylcholinesterase and cholinesterase activities in mice. *Experimental Neurology*, **22**, 493-503.

Lashley, K. S. (1929). *Brain mechanisms and intelligence*. Chicago: University of Chicago Press.

Leah, J., Allardyce, H., & Cummins, R. (1985). Evoked cortical potential correlates of rearing environment in rats. *Biological Psychology*, **20**, 21-29.

Levitsky, D. A., and Barnes, R. H. (1972). Nutritional and environmental interactions in the behavioral development of the rat: Long-term effects. *Science*, **176**, 68-71.

Lindroos, O. F. C., Riittinen, M.-L. A., Veilahti, J. V., Tarkkonen, L. J., Multanen, H. I., & Bergstrom, R. M. (1984). Overstimulation, occipital/somesthetic cerebral cortical depth, and cortical asymmetry in mice. *Developmental Psychobiology*, **17**, 547-554.

Lore, R. K. (1969). Pain avoidance behavior of rats reared in restricted and enriched environments. *Developmental Psychology*, **1**, 482-484.

Lore, R. K., & Levowitz, A. (1966). Differential rearing and free versus forced exploration. *Psychonomic Science*, **5**, 421-422.

Luchins, A. S., & Forgus, R. H. (1955). The effects of differential post-weaning environment on the rigidity of an animal's behavior. *The Journal of Genetic Psychology*, **86**, 51-58.

Luciano, D., & Lore, R. (1975). Aggression and social experience in domesticated rats. *Journal of Comparative and Physiological Psychology*, **88**, 917-923.

Lynch, G., McGaugh, J. L., & Weinberger, N. M., Editors (1984). *Neurobiology of Learning and Memory*. New York: Guilford.

Maki, W. S. (1971). Failure to replicate effect of visual pattern restriction on brain and behavior. *Nature (New Biology)*, **233**, 63-64.

Markowitz, H. (1982). *Behavioral Enrichment in the Zoo*. New York: Van Nostrand Reinhold.

Markowitz, H., & Spinelli, J. S. (1986). *The Interface Between Research and Environmental Enrichment*. Symposium paper presented at the annual meeting of the American Psychological Association, Washington, DC.

Matthies, H. (1986). *Learning and Memory: Mechanisms of Information Storage in the Nervous System. Advances in the Biosciences*, vol. 59. New York: Pergamon Press.

McCall, R. B., Lester, M. L., & Dolan, C. G. (1969). Differential rearing and the exploration of stimuli in the open field. *Developmental Psychology*, **1**, 750-762.

McGrath, M. J., & Cohen, D. B. (1978). REM sleep facilitation and adaptive waking behavior: A review of the literature. *Psychological Bulletin*, **85**, 24-57.

Mileusnic, R., Rose, S. P. R., & Tillson, P. (1980). Passive avoidance learning results in region-specific changes in concentration of and incorporation into colchicine-binding proteins in the chick forebrain. *Journal of Neurochemistry*, **34**, 1007-1015.

Mirmiran, M., Brenner, E., & Uylings, H. M. M. (1983). Noradrenaline depletion during early development inhibits cortical response to environmental experience. *Society for Neuroscience Abstracts*, **9**, 945.

Mirmiran, M., Van den Dungen, H., & Uylings, H. B. M. (1982). Sleep patterns during rearing under different environmental conditions in juvenile rats. *Brain Research*, **233**, 287-298.

Mishkin, M., Malamut, B., & Bachevalier, J. (1984). Memories and habits: two neural systems. In G. Lynch, J. L. McGaugh, and N. M. Weinberger (Eds.), *Neurobiology of Learning and Memory*, 65-77. New York: Guilford Press.

Mizumori, S. J. Y., Rosenzweig, M. R., & Bennett, E. L. (1985). Long-term working memory in the rat: Effects of hippocampally applied anisomycin. *Behavioral Neuroscience*, **99**, 220-232.

Mollgaard, K., Diamond, M. C., Bennett, E. L., Rosenzweig, M. R., & Lindner, B. (1971). Quantitative synaptic changes with differential experience in rat brain. *International Journal of Neuroscience*, **2**, 113-128.

Morgan, M. J. (1973). Effects of postweaning environment of learning in the rat. *Animal Behaviour*, **21**, 429-442.

Morris, R. (1984). Development of a water-maze procedure for studying spatial learning in the rat. *Journal of Neuroscience Methods*, **11**, 47-60.

Nyman, A. J. (1967). Problem solving in rats as a function of experience at different ages. *Journal of Genetic Psychology*, **110**, 31-39.

O'Keefe, J., & Nadel, L. (1978). *Hippocampus as a cognitive map*. New York: Oxford University Press.

O'Shea, L., Saari, M., Pappas, B. A., Ings, R., & Stange, K. (1983). Neonatal 6-hydroxydopamine attenuates the neural and behavioral effects of enriched rearing in the rat. *Society for Neuroscience Abstracts*, **9**, 558.

Ough, B. R., Beatty, W. W., & Khalili, J. (1972). Effects of isolated and enriched rearing on response inhibition. *Psychonomic Science*, **27**, 293-294.

Pappas, B. A., Saari, M., Smyth, J., O'Shea, L., Murtha, S., Stange K., & Ings, R. (1984). Neonatal forebrain norepinephrine loss eliminate rearing effects in the rat. *Society for Neuroscience Abstracts*, **10**, 1174.

Pappas, C. T., Diamond, M. C., & Johnson, R. E. (1978). Effects of ovariectomy and differential experience on rat cerebral cortical morphology. *Brain Research*, **154(1)**, 53-60.

Pearlman, C. (1983). Impairment of environmental effects on brain weight by adrenergic drugs in rats. *Physiology and Behavior*, **30**, 161-163.

Por, S. B., Bennett, E. L., & Bondy, S. C. (1982). Environmental enrichment and neurotransmitter receptors. *Behavioral and Neural Biology*, **34**, 132-140.

Pryor, R. (1964). *Unpublished doctoral dissertation*. University of California, Berkeley.

Quay, W. B., Bennett, E. L., Rosenzweig, M. R., & Krech, D. (1969). Effects of isolation and environmental complexity on brain and pineal organ. *Physiology and Behavior*, 4, 489-494.

Rabinovitch, M. S., & Rosvold, H. E. (1951). A closed-field intelligence test for rats. *Canadian Journal of Psychology*, **5**, 122-128.

Ramon y Cajal, S. (1894). La fine structure des centres nerveux. *Proceedings Royal Society London*, **55**, 444-468.

Ray, O. S., & Hochhauser, S. (1969). Growth hormone and environmental complexity effects on behavior in the rat. *Developmental Psychobiology*, 1, 311-317.

Reid, L. D., Gill, J. H., & Porter, P. B. (1968). Isolated rearing and Hebb-Williams maze performance. *Psychological Reports*, **22**, 1073-1077.

Renner, M. J. (1987a). Experience-dependent changes in exploratory behavior in the adult rat (*Rattus norvegicus*): Overall activity level and interactions with objects. *Journal of Comparative Psychology*, **101**(1), in press.

Renner, M. J. (1987b). *Learning during exploration: Acquisition of functionally significant information during spontaneous activity.* Manuscript submitted for publication.

Renner, M. J., Blank, C. L., Freeman, K., & Lin, L. (1986). Environmental enrichment and monoamine metabolism: gross levels, metabolites, and regional specificity. *Society for Neuroscience Abstracts*, **12**, 1136.

Renner, M. J., & Rosenzweig, M. R. (1983). *Do group- and enriched-housed rats differ in social interactions?.* Paper presented at the meeting of the American Psychological Association, Anaheim, California.

Renner, M. J., & Rosenzweig, M. R. (1986a). Social interactions among rats housed in grouped and enriched conditions. *Developmental Psychobiology*, **19**, 303-313.

Renner, M. J., & Rosenzweig, M. R. (1986b). Object interactions in juvenile rats (*Rattus norvegicus*): Effects of different experiential histories. *Journal of Comparative Psychology*, **100**, 229-236.

Renner, M. J., & Rosenzweig, M. R. (1987). The Golden-mantled ground squirrel (*Spermophilus lateralis*) as a model for the effects of environmental enrichment in solitary animals. *Developmental Psychobiology*, **20**, in press.

Renner, M. J., Rosenzweig, M. R., Bennett, E. L., & Alberti, M. (1981). Progressive increases in environmental complexity produce monotonic changes in brain and behavior. *Society for Neuroscience Abstracts*, **7**, 842.

Riege, W. H. (1971). Environmental influences on brain and behavior of year-old rats. *Developmental Psychobiology*, **4**, 157-167.

Riege, W. H., & Morimoto, H. (1970). Effects of chronic stress and differential environments upon brain weights and biogenic amine levels in rats. *Journal of Comparative and Physiological Psychology*, **71**, 396-404.

Robbins, S. F. (1977). The effects of differential rearing environments on preference for complexity in the rat. *Dissertation Abstracts International*, **38**(5B), 2405-2406.

Roeder, J.-J., Chetcuti, Y., & Will, B. (1980). Behavior and length of survival of populations of enriched and impoverished rats in the presence of a predator. *Biology of Behavior*, **5**, 361-369.

Rose, F. D., Love, S., & Dell, P. A. (1986). Differential reinforcements effects in rats reared in enriched and impoverished environments. *Physiology and Behavior*, **36**, 1139-1145.

Rose, S. P. R., & Harding, S. (1984). Training increases [^{3}H]fucose incorporation in chick brain only if followed by memory storage. *Neuroscience*, **12**, 663-667.

Rosenzweig, M. R., & Bennett, E. L. (1969). Effects of differential environments on brain weights and enzyme activities in gerbils, rats and mice. *Developmental Psychobiology*, **2**, 87-95.

Rosenzweig, M. R., & Bennett, E. L. (1972). Cerebral changes in rats exposed individually to an enriched environment. *Journal of Comparative and Physiological Psychology*, **80**, 304-313.

Rosenzweig, M. R., & Bennett, E. L. (1977). Effects of environmental enrichment and impoverishment on learning and on brain values in rodents. In A. Oliverio, (Ed.), *Genetics, Environment and Intelligence*, 163-196. Amsterdam: Elsevier North-Holland.

Rosenzweig, M. R., & Bennett, E. L. (1978). Experiential influences on brain anatomy and brain chemistry in rodents. In G. Gottlieb (Ed.), *Studies on the Development of Behavior and the Nervous System*, 289-327. New York: Academic Press.

Rosenzweig, M. R., & Bennett, E. L. (1980). How plastic is the nervous system?. In Almi and S. Finger (Eds.), *A Comprehensive Handbook of Behavioral Medicine*, 149-185. Spectrum Publications.

Rosenzweig, M. R., & Bennett, E. L. (1984). Basic processes and modulatory influences in the stages of memory formation. In G. Lynch, J. L. McGaugh, and N. M. Weinberger (Eds.), *Neurobiology of Learning and Memory*, 263-288.

Rosenzweig, M. R., Bennett, E. L., & Alberti, M. (1984). Multiple effects of lesions on brain structure in young rats. In S. Finger and D. G. Stein (Eds.), *Early Brain Damage*, Volume 2, 49-70. New York: Academic Press.

Rosenzweig, M. R., Bennett, E. L., Alberti, M., Morimoto, H., & Renner, M. J. (1982). Effects of differential environments and hibernation on ground squirrel brain measures. *Society for Neuroscience Abstracts*, **8**, 669.

Rosenzweig, M. R., Bennett, E. L., & Diamond, M. C. (1972a). Cerebral effects of differential experience in hypophysectomized rats. *Journal of Comparative and Physiological Psychology*, **79**, 56-66.

Rosenzweig, M. R., Bennett, E. L., & Diamond, M. C. (1972b). Chemical and anatomical plasticity of brain: Replications and extensions. In: J. Gaito (Ed.), *Macromolecules and Behavior*, 2nd Ed., 205-277.

Rosenzweig, M. R., Bennett, E. L., & Diamond, M. C. (1972c). Brain changes in response to experience. *Scientific American*, **226**(2), 22-29.

Rosenzweig, M. R., Bennett, E. L., Diamond, M. C., Wu, S.-Y., Slagle, R. W., & Saffran, E. (1969). Influences of environmental complexity and visual stimulation on development of occipital cortex in rat. *Brain Research*, **14**, 427-445.

Rosenzweig, M. R., Bennett, E. L., Hebert, M., & Morimoto H. (1978). Social grouping cannot account for cerebral effects of enriched environments. *Brain Research*, **153**, 563-576.

Rosenzweig, M. R., Bennett, E. L., & Krech, D. (1964). Cerebral effects of environmental complexity and training among adult rats. *Journal of Comparative and Physiological Psychology*, **57**, 438-439.

Rosenzweig, M. R., Bennett, E. L., Renner, M. J., & Alberti, M. (1987). *Laboratory enrichment and the natural baseline: Effects of differential environments on brain measures in two species of ground squirrels*. Unpublished manuscript.

Rosenzweig, M. R., Bennett, E. L., & Sherman, P. W. (1980). Effects of hibernation and differential environments on weights and nucleic acids in brain of Belding's ground squirrels. *Society for Neuroscience Abstracts*, **6**, 635.

Rosenzweig, M. R., Krech, D., & Bennett, E. L. (1960). A search for relations between brain chemistry and behavior. *Psychological Bulletin*, **57**, 476-492.

Rosenzweig, M. R., Krech, D., Bennett, E. L., & Diamond, M. C. (1962). Effects of environmental complexity and training on brain chemistry and anatomy: A replication and extension. *Journal of Comparative and Physiological Psychology*, **55**, 429-437.

Rosenzweig, M. R., Krech, D., Bennett, E. L., & Zolman, J. F. (1962). Variation in environmental complexity and brain measures. *Journal of Comparative and Physiological Psychology*, **55**, 1092-1095.

Rosenzweig, M. R., Love, W., & Bennett, E. L. (1968). Effects of a few hours a day of enriched experience on brain chemistry and brain weights. *Physiology and Behavior*, **3**, 819-825.

Royce, J. R. (1977). On the construct validity of open-field measures. *Psychological Bulletin*, **84**, 1098-1106.

Sackett, G. P. (1972). Exploratory behavior of rhesus monkeys as a function of rearing experiences and sex. *Developmental Psychology*, **6**, 260-270.

Sandman, C. A., & Donnelly, J. (1983). Age differences in P300 and its relationship to activity in the elderly. *Psychophysiology*, **20**, 467.

Sara, V. R., King, T. L., & Lazarus, L. (1976). The influence of early nutrition and environmental rearing on brain growth and behavior. *Experientia*, **32**(12), 1538-1540.

Schliebs, R., Rose, S. P. R., & Stewart, M. G. (1985). Effect of passive avoidance training on *in vitro* protein synthesis in forebrain slices of day-old chicks. *Journal of Neurochemistry*, **44**, 1014-1028.

Schwartz, S. (1964). Effects of neonatal cortical lesions and early environmental factors on adult rat behavior. *Journal of Comparative and Physiological Psychology*, **57**, 72-77.

Seligman, M. E. P. (1970). On the generality of the laws of learning. *Psychological Review*, **77**, 406-418.

Shapiro, M. L., Nilsson, O., Gage, F. H., Olton, D. S., & Bjorklund, A. (1986). Fetal basal forebrain transplants to the hippocampus restore working memory in rats with fimbria-fornix lesions. *Society for Neuroscience Abstracts*, **12**, 742.

Sharp, P. E., Barnes, C. A., & McNaughton, B. L. (1985). Age-related differences in the decay of environmentally-induced hippocampal plasticity. *Society for Neuroscience Abstracts*, **11**, 721.

Sharp, P. E., McNaughton, B. L., & Barnes, C. A. (1983). Spontaneous synaptic enhancement in hippocampi of rats exposed to a spatially complex environment. *Society for Neuroscience Abstracts*, **9**, 647.

Sharp, P. E., McNaughton, B. L., & Barnes, C. A. (1985). Enhancement of hippocampal field potentials in rats exposed to a novel, complex environment. *Brain Research*, **339**, 361-365.

Sholl, D. A. (1956). *The organization of the cerebral cortex*. London: Methuen.

Singh, D., Johnston, R. J., & Klosterman, H. J. (1967). Effect on brain enzyme and behavior in the rat of visual pattern restriction in early life. *Nature*, **216**, 1337-1338.

Singh, D., Johnston, R. J., & Klosterman, H. J. (1970). Effect of visual pattern restriction in early life on brain enzyme in the rat. *Psychonomic Science*, **19**, 173-174.

Singh, D., Johnston, R. J., & Maki, W. S. (1969). Effects of visual pattern restriction on sensory reinforcement in the rat. *Psychonomic Science*, **15**, 117-118.

Singh, D., Maki, W. S., Johnston, R. J., & Klosterman, H. J. (1970). Effects of visual pattern restriction in early life on brain enzyme, body weight and learning in the rat. *Nature*, **228**, 471-472.

Sirevaag, A. M., & Greenough, W. T. (1985a). Differential rearing effects on rat visual cortex synapses. II. Synaptic Morphometry. *Developmental Brain Research*, **19**, 215-226.

Sirevaag, A. M., & Greenough, W. T. (1985b). Numerical density of boutons and its relationship to neuronal, axonal and dendritic density in the occipital cortex of rats reared in complex, social and isolated environments. *Society for Neuroscience Abstracts*, **11**, 99.

Sirevaag, A. M., & Greenough, W. T. (1986). Multivariate analyses of morphological measurements can distinguish among rats reared in complex, social or individual environments. *Society for Neuroscience Abstracts*, **12**, 1284.

Sjoden, P.-O. (1976). Effects of neonatal thyroid hormone stimulation and differential preweaning rearing on spatial discrimination learning in rats. *Physiological Psychology*, **4**, 515-520.

Smith, H. V. (1972). Effects of environmental enrichment on open-field activity and Hebb-Williams problem-solving in rats. *Journal of Comparative and Physiological Psychology*, **80**, 163-168.

Squire, L. R., & Butters, N. (Eds.) (1984). *The Neuropsychology of Memory*. New York: Guilford Press.

Stein, D. G., Finger, S., & Hart, T. (1983). Brain damage and recovery: problems and perspectives. *Behavioral and Neural Biology*, **37**, 185-222.

Studelska, D. R., & Kemble, E. D. (1979). Effects of briefly experienced environmental complexity on open-field behavior in rats. *Behavioral and Neural Biology*, **26**, 492-496.

Sturgeon, R. D., & Reid, L. D. (1971). Rearing variations and Hebb-Williams maze performance. *Psychological Reports*, **29**, 571-580.

Suarez, S. D., & Gallup, G. G. (1981). Predatory overtones of open-field testing in chickens. *Animal Learning and Behavior*, **9**, 153-163.

Sukumar, R., Rose, S. P. R., & Burgoyne, (1980). Increased incorporation of [^{3}H]fucose into chick brain glycoproteins following training on a passive avoidance task. *Journal of Neurochemistry*, **34**, 1000-1006.

Susser, E. R., & Wallace, R. B. (1982). The effects of environmental complexity on the hippocampal formation of the adult rat. *Acta Neurobiologica Experimentalis*, **42**, 203-207.

Szeligo, F. (1977). Quantitative differences in oligodendrocytes and myelinated axons in the brains of rats raised in enriched, control, and impoverished environments. *Anatomical Record*, **187**, 726-727.

Szeligo, F., & Leblond, C. P. (1977). Response of the three main types of glial cells of cortex and corpus callosum in rats handled during suckling or exposed to enriched, control, or impoverished environments following weaning. *Journal of Comparative Neurology*, **172**, 247-264.

Tagney, J. (1973). Sleep patterns related to rearing rats in enriched and impoverished environments. *Brain Research*, **53**, 353-361.

Teyler, T. J., & DiScenna, P. (1986). The hippocampal memory indexing theory. *Behavioral Neuroscience*, **101**, 147-154.

Thompson, R. F. (1986). The neurobiology of learning and memory. *Science*, **233**, 941-947.

Tryon, R. C. (1940). Genetic differences in maze-learning ability in rats. *Yearbook of the National Society for the Study of Education*, **39**, 111-119.

Turner, A. M., & Greenough, W. T. (1983). Numerical density of synapses in neuropil of occipital cortex of rats reared in complex, social, or isolated environments. *Society for Neuroscience Abstracts*, **9**, 55.

Turner, A. M., & Greenough, W. T. (1985). Differential rearing effects on rat visual cortex synapses. I. Synaptic and neuronal density and synapses per neuron. *Brain Research*, **329**, 195-203.

Uphouse, L. (1978). In vitro RNA synthesis by chromatin from three brain regions of differentially reared rats. *Behavioral Biology*, **22**, 39-49.

Uphouse, L. (1980). Reevaluation of mechanisms that mediate brain differences between enriched and impoverished animals. *Psychological Bulletin*, **88**, 215-232.

Uphouse, L., & Brown, H. (1981). Effect of differential rearing on brain, liver, and adrenal tissues. *Developmental Psychobiology*, **14**, 273-278.

Uphouse, L., & Tedeschi, B. (1979). Environmental enrichment and brain chromatin. *Behavioral and Neural Biology*, **25**, 268-270.

Uylings, H., Kuypers, K., Diamond, M., & Veltman W. (1978). Effects of differential environments on plasticity of dendrites of cortical pyramidal neurons in adult rats. *Experimental Neurology*, **62**, 658-677.

Van Woerden, G. J. M. (1986). *Effects of Differential Experience on Brain and Behaviour in the Rat*. Unpublished doctoral dissertation, Universiteit te Nijmegen, The Netherlands.

Volkmar, F. R., & Greenough, W. T. (1972). Rearing complexity affects branching of dendrites in the visual cortex of the rat. *Science*, **176**, 1447-1449.

Wallace, C. S., Black, J. E., Hwang, H. M., & Greenough W. T. (1986). Housing complexity affects body and adrenal weight of adult rats within 10 days exposure. *Society for Neuroscience Abstracts*, **12**, 1283.

Walsh, R. N. (1980). Effects of environmental complexity and deprivation on brain chemistry and Physiological: a review. *International Journal of Neuroscience*, **11**, 77-89.

Walsh, R. N. (1981). *Towards an Ecology of Brain*. New York: SP Medical and Scientific Books.

Walsh, R. N., Budtz-Olsen, O. E., Penny, J. E., & Cummins, R. A. (1969). The effects of environmental complexity on the histology of the rat hippocampus. *Journal of Comparative Neurology*, **137**, 361-366.

Walsh, R. N., Budtz-Olsen O. E., Torok. A., & Cummins, R. A. (1971). Environmentally induced changes in the dimensions of the rat cerebrum. *Developmental Psychobiology*, **4**, 115-122.

Walsh, R. N., & Cummins, R. A. (1976a). Electron microscopic observations of occipital cortex in differentially reared rats. *Society for Neuroscience Abstracts*, **2**, 839.

Walsh, R. N., & Cummins, R. A. (1976b). The open-field test: A critical review. *Psychological Bulletin*, **83**, 482-504.

Walsh, R. N., & Cummins, R. A. (1978). Caveats for future research on the open-field test: comment on Royce. *Psychological Bulletin*, **85**, 587-589.

Walsh, R. N., & Cummins, R. A. (1979). Changes in hippocampal neuronal nuclei in response to environmental stimulation. *International Journal of Neuroscience*, **9**, 209-212.

Walsh, R. N., Cummins, R. A., & Budtz-Olsen, O. E. (1973). Environmentally induced changes in the rat cerebrum: a replication and extension. *Developmental Psychobiology*, **6**, 3-7.

Walsh, R. N., & Greenough, W. T., Editors (1976). *Environ-*

ments as Therapy for Brain Dysfunction. New York: Plenum Press.

Warren, J. M. (1985). Environmental enrichment and learning by DBA/2J mice. *Festschrift for M. M. Sinha*, in press.

Warren, J. M., Zerweck, C., & Anthony, A. (1982). Effects of environmental enrichment on old mice. *Developmental Psychobiology*, **15**, 13-18.

Weinberger, N. M., McGaugh, J. L., & Lynch, G. (1985). *Memory Systems of the Brain.* New York: Guilford.

Welch, B. L., Brown, D. G., Welch, A. S., & Lin, D. C. (1974). Isolation, restrictive confinement or crowding of rats for one year. I. Weight, nucleic acids and protein of brain regions. *Brain Research*, **75**, 71-84.

Welker, W. I. (1957). "Free" versus "forced" exploration of a novel situation by rats. *Psychological Reports*, **3**, 95-100.

Wenzel, J., Kammerer, E., Frotscher, M., Joschko, R., Joschko, M., & Kaufmann, W. (1977a). Elektronenmikroskopische und morphometrische Untersuchungen an Synapsen des Hippocampus nach Lernexperimenten bei der Ratte. *Zeitschrift fur mikroskopische-anatomische Forschung*, **91**, 74-93. Leipzig.

Wenzel, J., Kammerer, E., Joschko, R., Joschko, M., Kaufmann, W., Kirsche, W., & Matthies, H. (1977b). Der Einfluss eines Lernexperimentes auf de Synapsenanzahl im Hippocampus der Ratte. Elektronenmikroskopische und morphometrische Untersuchungen. *Zeitschrift fur mikroskopische-anatomische Forschung*, **91**, 57-73. Leipzig.

Wesa, J. M., Chang, F.-L. F., Greenough, W. T., & West, R. W. (1982). Synaptic contact curvature: Effects of differential rearing on rat occipital cortex. *Developmental Brain Research*, **4**, 253-257.

West, R. W., & Greenough, W. T. (1972). Effects of environmental complexity on cortical synapses of rats: Preliminary results. *Behavioral Biology*, **7**, 279-283.

Whimbey, A. E., & Denenberg, V. H. (1967). Two independent dimensions in open-field performance. *Journal of Comparative and Physiological Psychology*, **63**, 500-504.

Whishaw, I. Q., Zaborowski, J-A., & Kolb, B. (1984). Postsurgical enrichment aids adult hemidecorticated rats on a spatial navigation task. *Behavioral and Neural Biology*, **42**, 183-190.

White, L. (1975). *Behavioral response of San Francisco Zoo's orangutans to environmental enrichment.* Paper presented at the meeting of the Western Psychological Association, San Francisco.

Will, B. E., & Rosenzweig, M. R. (1976). Effets de l'environment sur la recuperation fonctionnelle apres lesions cerebrales chez les rats adultes. *Biology of Behavior*, **1**, 5.

Will, B., Kelche, C., & Dalrymple-Alford, J. C. (1986). Recovery of function after a complete fimbria-fornix lesion: interaction of fetal septal transplants and post-operative. *Society for Neuroscience Abstracts*, **12**, 975.

Will, B., Kelche, C., & Deluzarche, F. (1981). Effects of postoperative environment on functional recovery after entorhinal lesions in the rat. *Behavioral and Neural Biology*, **33**, 303-316.

Will, B., & Kelche, C. R. (1979). Effects of different postoperative environments on the avoidance behavior of rats with hippocampal lesions: Recovery or improvement of function?. *Behavioral and Neural Biology*, **27**, 96-106.

Will, B., Rosenzweig, M. R., & Bennett, E. L. (1976). Effects of differential environments on recovery from neonatal brain lesions, measured by problem solving scores and brain dimensions. *Physiology and Behavior*, **16**, 603-611.

Will, B., Rosenzweig, M. R., Bennett, E. L., Hebert, M., & Morimoto, H. (1977). Relatively brief environmental enrichment aids recovery of learning capacity and alters brain measures after postweaning brain lesions in rats. *Journal of Comparative and Physiological Psychology*, **91**, 33-50.

Will, B., Sutter, A. R., & Offerlin, M. R. (1977). Effects of methamphetamine and enriched experience on behavioral recovery after brain damage. *Psychopharmacology*, **51**, 273-277.

Wilson, M., Warren, J. M., & Abbott, L. (1965). Infantile stimulation, activity and learning by cats. *Child Development*, **36**, 843-853.

Wood-Gush, D., Stolba, A., & Miller, C. (1983). Exploration in farm animals and animal husbandry. In J. Archer and L. I. A. Birke (Eds.), *Exploration in Animals and Humans*, 198-209. New York: Van Nostrand Reinhold.

Woods, P. J. (1959). The effects of free and restricted environmental experience on problem-solving behavior in the rat. *Journal of Comparative and Physiological Psychology*, **52**, 399-402.

Woods, P. J., Fiske, A. S., & Ruckelshaus, S. I. (1961). The effects of drives conflicting with exploration on the problem-solving behavior of rats reared in free and restricted environments. *Journal of Comparative and Physiological Psychology*, **54**, 167-169.

Woods, P. J., Ruckelshaus, S. I., & Bowling, D. M. (1960). Some effects of "free" and "restricted" environmental rearing conditions upon adult behavior in the rat. *Psychological Reports*, **6**, 191-200.

Zimbardo, P. G., & Montgomery, K. C. (1957). Effects of "free - environment" rearing upon exploratory behavior. *Psychological Reports*, **3**, 589-594.

Zolman, J. F., & Morimoto, H. (1965). Cerebral changes related to duration of environmental complexity and locomotor activity. *Journal of Comparative and Physiological Psychology*, **60**, 382-387.

Zubeck, J. P. (1951). Recent electrophysiological studies of the cerebral cortex: Implications for localization of sensory function. *Canadian Journal of Psychology*, **5**, 110-121.

Index of Names